James T. Luxon
David E. Parker

GMI Engineering & Management Institute
Flint, Michigan 48502–2276

INDUSTRIAL LASERS AND THEIR APPLICATIONS

PRENTICE-HALL, INC., Englewood Cliffs, New Jersey 07632

Library of Congress Cataloging in Publication Data

LUXON, JAMES T. (date)
  Industrial lasers and their applications.

  Includes bibliographies and index.
  1. Lasers—Industrial applications. I. Parker,
David E. II. Title.
TA1677.L89 1984   621.36′6   84–13385
ISBN 0–13–461369–4

To our wives, Sally and Nancy

Editorial/production supervision and
   interior design: *Gretchen Chenenko and Nancy DeWolfe*
Cover design: *Whitman Studio, Inc.*
Manufacturing buyer: *Gordon Osbourne*

Printed in the United States of America

10 9 8 7 6 5 4 3 2

ISBN 0-13-461369-4   01

PRENTICE-HALL INTERNATIONAL, INC., *London*
PRENTICE-HALL OF AUSTRALIA PTY. LIMITED, *Sydney*
EDITORA PRENTICE-HALL DO BRASIL, LTDA., *Rio de Janeiro*
PRENTICE-HALL CANADA INC., *Toronto*
PRENTICE-HALL OF INDIA PRIVATE LIMITED, *New Delhi*
PRENTICE-HALL OF JAPAN, INC., *Tokyo*
PRENTICE-HALL OF SOUTHEAST ASIA PTE. LTD., *Singapore*
WHITEHALL BOOKS LIMITED, *Wellington, New Zealand*

Contents

Preface

The purpose of this book is to provide the reader with an introduction to lasers and their industrial applications. To facilitate this objective, such devices as photodetectors and modulators, which are frequently found in laser applications, are also covered. And to make the book as self-contained as possible, the concepts of basic optics that are pertinent to lasers and their applications are presented. Many engineering students do not cover this material formally in their course work; moreover, many working engineers and scientists either have not had training in optics or have been away from it for a long time and may need a refresher. Some laser theory is presented to provide a working understanding of the laser and to clear away the mysticism surrounding the device. When tools are understood, they are used more frequently and used properly.

The topic of laser beam optics, including propagation, focusing, and depth of focus, is covered in some detail for both Gaussian and higher-order mode beams because such information is of practical value to industrial applications of lasers.

A chapter on optical detectors, including detector arrays, is preceded by a short chapter on semiconductors, to enhance the understanding of solid-state optical devices, and by a chapter in which radiometry, photometry, and optical device parameters are discussed.

It would be impossible to present an exhaustive treatment of the interaction of high-power laser beams and matter, but some of the most pertinent cases are presented in a chapter on laser beam materials interaction. Separate chapters are devoted to industrial applications of low-power and high-power lasers. Specific types of applications are presented along with additional theoretical or conceptual material where required.

This is not a book on lasers but rather a book that is intended to help prepare engineering students or practicing engineers and scientists for the practical application of lasers in an industrial manufacturing setting. Thus the book may be used for a one-semester, junior–senior level course on lasers and laser applications or by practicing engineers and scientists who need to learn quickly the essentials of lasers and their applications. Greater depth in the topics on lasers or materials interactions, for example, can then be obtained from many more advanced books.

This book can be used in several different ways. The first chapter on basic optics can be omitted if the reader has a background in optics. The chapters on semiconductors, parameters, radiometry, and devices can be omitted if the reader's interests do not lie in these areas; these topics are not essential to the remainder of the book.

Readers who have some familiarity with lasers and their properties can omit the overview chapter on lasers. Any chapters on low-power or high-power applications may be omitted without loss of continuity.

The authors are greatly indebted to a number of people. We want to express our appreciation to our families for their patience and encouragement. We would like to thank many of our students who gave us constructive criticism and other assistance. We would particularly like to thank Mark Sparschu and Jim McKinley for reading Chapters 9 and 10 and working the problems. We also want to thank Ms. Barbara Parker for her skillful proofreading and Ms. Judy Wing for her patience, skill, and good humor in typing much of the manuscript.

JAMES T. LUXON
DAVID E. PARKER

chapter 1

Principles
of Optics

This chapter is intended to provide the reader with a basic working knowledge of the principles of optics, including a description of the nature of electromagnetic radiation as well as geometrical and physical optics. This chapter also provides a basis for much of what follows in subsequent chapters.

1-1 NATURE OF ELECTROMAGNETIC RADIATION

Electromagnetic radiation exhibits both wavelike and particlelike characteristics, as does matter when it comes in small enough packages, like electrons. Both aspects of electromagnetic radiation are discussed and both are relevant to understanding lasers. From the point of view of its wavelike characteristics, electromagnetic radiation is known to exhibit wavelengths from less than 10^{-13} m to over 10^{15} m. Included in this range, in order of increasing wavelength, are gamma rays, x rays, ultraviolet waves (uv), visible light, infrared (ir) light, microwaves, radio waves, and power transmission waves. Figure 1–1 illustrates the various parts of the range of electromagnetic waves as a function of wavelength.

The sources of gamma rays (γ rays) are nuclear transitions involving radioactive decay. X rays are produced through electronic transitions deep in the electronic structure of the atom. Ultraviolet waves result from electronic transitions involving fairly high energy and overlap the x-ray region somewhat. Visible radiation extends from about 0.35 μm to 0.75 μm and is due to electronic transitions, primarily of valence electrons. Infrared radiation results from electronic transitions at the near visible end and molecular vibrations toward the

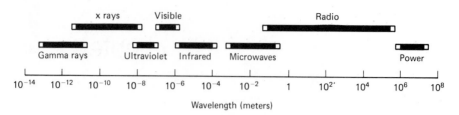

Figure 1–1 Electromagnetic radiation as a function of wavelength.

long wavelength end. Microwaves and radio waves are produced by various types of electronic oscillators and antennas, respectively.

The term *light* is used loosely to refer to radiation from uv through ir.

The wavelike properties of electromagnetic radiation can be deduced from the wave equation presented here in one-dimensional form

$$\frac{\partial^2 E}{\partial z^2} = \frac{1}{c^2} \frac{\partial^2 E}{\partial t^2} \tag{1-1}$$

where c is the velocity of light and E is the electric field intensity. The wave equation can be derived from Maxwell's equations, the foundation of all classical electromagnetic theory. The symbol E in Eq. (1–1) may represent any one of the various electromagnetic field quantities, but for our purposes the electric field intensity is of greatest interest.

Another relationship that can be deduced from Maxwell's equations that is of use to us is Poynting's theorem

$$\mathbf{S} = \mathbf{E} \times \mathbf{H} \tag{1-2}$$

where $\mathbf{S}$ is power flow per unit area, $\mathbf{E}$ is electric field intensity, and $\mathbf{H}$ is magnetic field intensity. For a freely propagating electromagnetic wave, it reduces to

$$S_{\text{ave}} = \tfrac{1}{2} EH \tag{1-3}$$

where S_{ave} is the average power flow per unit area and E and H are amplitudes.

Light may be thought of as being composed of sinusoidal components of electric and magnetic fields from the point of view that electromagnetic radiation is a wave. For a simple electromagnetic wave propagating in an unbounded medium (the electric field varying parallel to a single direction, referred to as *linear polarization*), the wave may be schematically represented as in Fig. 1–2.

The electric and magnetic fields are oriented at right angles to each other and to the direction of propagation z. $\mathbf{E}$, $\mathbf{H}$, and z form a right-hand triad; that is, $\mathbf{E} \times \mathbf{H}$ gives the direction of propagation. Ordinary (unpolarized) light contains a mixture of polarizations in all directions perpendicular to the direction of propagation. Because of the vector nature of the electric field, unpolarized light can be thought as an equal mix of electric field strength in orthogonal

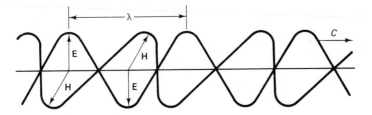

Figure 1-2   Propagation of a plane-polarized electromagnetic wave.

directions, say x and y, with random phase relations between the various contributions to the electric field. The significance of this statement will become apparent later on.

The speed of propagation in free space (vacuum) is approximately 3×10^8 m/s and is equal to $1/\sqrt{\mu_0 \epsilon_0}$ according to classical electromagnetic wave theory. For an electromagnetic wave propagating in a dielectric medium, the speed is

$$\frac{1}{\sqrt{\mu \epsilon}} = \frac{1}{\sqrt{\mu_r \mu_0 \epsilon_r \epsilon_0}} = \frac{c}{\sqrt{\mu_r \epsilon_r}}$$

where c is the speed of light in free space and μ_r and ϵ_r are the relative permeability and permittivity of the medium, respectively. In nonmagnetic materials $v = c/\sqrt{\epsilon_r}$. The refractive index of a dielectric medium is defined by

$$n = \frac{c}{v} \tag{1-4}$$

and so it is seen that $n = \sqrt{\epsilon_r}$ for most dielectrics.

It is possible to show that $E = ZH$, where Z is called the intrinsic impedance of the medium. This fact can be used to put Eq. 1-3 in the form

$$I = \frac{1}{2} \frac{E^2}{Z} \tag{1-5}$$

where I is the irradiance (power per unit area). You may recognize the similarity between Eq. (1-5) and the equation for Joule heating in a resistor, which is $P = V^2/R$, where P is power, V is voltage, and R is resistance. The one-half does not appear in the Joule's law heating equation because V is a root mean square rather than an amplitude.

The intrinsic impedance of an unbounded dielectric is $Z = \sqrt{\mu/\epsilon}$, where $\sqrt{\mu_0/\epsilon_0}$, the intrinsic impedance of free space, is $377\,\Omega$. Then

$$Z = \frac{377\,\Omega}{\sqrt{\epsilon_r}} = \frac{377\,\Omega}{n} \tag{1-6}$$

for nonmagnetic dielectrics. Equation (1-5) can therefore be written

$$I = \frac{1}{2} \frac{E^2 n}{377\Omega} \tag{1-7}$$

The subject of electromagnetic wave propagation in conductors is beyond the scope of this book, but a few pertinent facts can be pointed out. The impedance of a good conductor is given by

$$Z = \frac{\omega}{\sigma} e^{-j(\pi/4)} \tag{1-8}$$

where ω is the radian frequency of the light and σ is the conductivity of the conductor. As can be seen from Eq. (1-8), Z is a complex impedance. E is very small in a conductor; H is large. When an electromagnetic wave strikes a conductor, E will go nearly to zero and H becomes large due to large induced surface currents. The results are considerable reflectance of the incident wave and rapid attenuation of the transmitted wave. The skin depth, which is a measure of how far the wave penetrates, is given by

$$\delta = \frac{1}{\sqrt{\pi\mu\sigma f}} \tag{1-9}$$

For frequencies of interest in this book, chiefly visible and ir to 10.6 μm, δ is extremely small and absorption can be assumed to take place at the surface for all practical purposes.

Based on the Drude free electron theory of metals, it can be shown that the fraction of the incident power absorbed by a metal is given approximately by

$$A = 4\sqrt{\frac{\pi c \epsilon_0}{\lambda \sigma}} \tag{1-10}$$

The reflectance is $R = 1 - A$, which, for copper with $\sigma = 5.8 \times 10^7 (\Omega - m)^{-1}$ at $\lambda = 10.6$ μm, leads to $R = 0.985$. Actual reflectances may exceed this value for very pure copper. For highly polished or diamond-turned copper mirrors the reflectance exceeds 0.99.

The particlelike behavior of electromagnetic radiation is exhibited in many experiments, such as the photoelectric effect and Compton scattering. It was in an explanation of the photoelectric effect in 1905 that Einstein proposed that electromagnetic radiation (light for short) is composed of bundles of energy, quanta, which are referred to as *photons*. The energy of each of these photons, he argued, is hf, where h is Planck's constant ($h = 6.6 \times 10^{-34}$ J $\cdot$ s). Planck had determined this constant previously in explaining the dependence of black-body radiation on frequency.

In 1924 DeBroglie proposed a mathematical model for the photon. This model consists of an infinite sum of waves of different frequencies within a finite frequency range with an appropriate amplitude function. It was really nothing more than a Fourier integral representation of a finite pulse, with the amplitude function chosen to produce a wave packet with minimum uncertainty products. A schematic representation of such a pulse is given in Fig. 1-3. The outer solid lines form an envelope of the amplitude of the actual wave. The uncertainties referred to concern the length of the packet L, its relation to the uncertainty in its momentum Δp, and the relation between frequency band-width and the time it takes the photon to pass a given point. These relations are

$$\Delta p L \leq \frac{h}{2\pi}$$

(1-11)

$$\Delta f \Delta t \leq \frac{1}{2\pi}$$

These relations will be useful later on in discussing coherence of light sources.

DeBroglie proposed a duality of both light and matter; that is, he suggested that matter should exhibit both wave and particle characteristics. It was later shown that electrons can be diffracted by crystals and the observed wavelength agreed with that predicted by DeBroglie. The DeBroglie wavelength can be deduced by setting Einstein's famous mass energy relationship $E = mc^2$ equal to the energy of a photon hf. Thus

$$E = mc^2 = hf$$

and $mc = hf/c$ is the momentum of a photon, and therefore momentum and wavelength are related by

$$p = mc = \frac{h}{\lambda}$$

(1-12)

This relation holds for both light and matter waves. These results can be used to show (left as an exercise for the student) that $dp/dt = P/c$ for a total absorption at a surface, where P is total power in the beam and c is the speed of light. The radiation pressure is dp/dt divided by the area of incidence. In general, this pressure can be written

$$\frac{F}{A} = \frac{(1 + R)P}{cA}$$

(1-13)

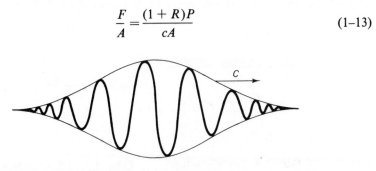

Figure 1-3 Schematic representation of mathematical model of a photon.

where R is the fraction of the incident power reflected and A is the area of incidence. This pressure can be substantial for a focused laser beam.

1-2 REFLECTION AND REFRACTION

The *law of reflection* states that, for specular reflection of light, the angle of incidence equals the angle of reflection. This situation is illustrated in Fig. 1-4. The reflected ray lies in the same plane as the incident ray and the normal to the surface. This plane is referred to as the *plane of incidence*. A specular surface is one with a surface finish characterized by rms variations in height and separation of peaks and valleys (surface roughness) much less than the wavelength of the light. In other words, a surface that is not a good specular surface in the visible could be quite specular at longer wavelengths. This is an important point to remember when working around high power—long wavelength lasers.

Most surfaces cause reflected light to contain a portion of specular and diffusely reflected (scattered) light. The diffuse reflection is the result of random reflections in all directions due to roughness of surface finish. An ideal diffuser scatters equal amounts of power per unit area per unit solid angle in all directions. Hence a perfect diffusing surface is equally bright from all viewing angles. Few surfaces approach the ideal case.

The law of refraction, or *Snell's law* as it is commonly called, is given by Eq. (1-14) and the angles are defined in Fig. 1-5.

$$n_1 \sin \theta_1 = n_2 \sin \theta_2 \qquad (1-14)$$

The law of reflection applies equally to all materials whereas Snell's law, in the form given in Eq. (1-14), is valid only for an interface between two dielectrics.

Two phenomena of importance relate simply to Snell's law. The first, total internal reflection (TIR), occurs when light travels from a medium of higher refractive index into one of the lower refractive indices. Because $n_2 < n_1$, Snell's law requires that $\theta_2 > \theta_1$. At an angle of incidence called the critical angle θ_2 becomes 90°. At this angle of incidence θ_c, and for all $\theta_1 > \theta_c$, all

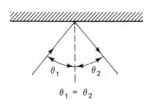

Figure 1-4 Law of reflection.

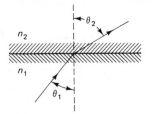

Figure 1-5 Refraction at a dielectric interface.

incident power is reflected. The critical angle is deduced from Snell's law to be

$$\theta_c = \sin^{-1} \frac{n_2}{n_1} \qquad (1\text{-}15)$$

The second phenomenon is Brewster's law, which has significance in many laser designs as well as other areas. Brewster's law states that when the reflected and refracted rays are at right angles to each other, the reflected light is linearly polarized perpendicular to the plane of incidence. The angle of incidence at which this occurs is called the Brewster angle θ_B. Figure 1-6 illustrates this phenomenon. Because θ_2 is the complement of θ_B, Snell's law gives

$$n_1 \sin \theta_B = n_2 \cos \theta_B \qquad (1\text{-}16)$$

or

$$\theta_B = \tan^{-1} \left(\frac{n_2}{n_1} \right)$$

This is a reversible phenomenon, unlike TIR, and the Brewster angle from side 2 to 1 is $90° - \theta_B$ or $\tan^{-1}(n_1/n_2)$.

When light is incident on a surface, a certain fraction of the light is absorbed or transmitted and the remainder is reflected. The fraction of power reflected, called the reflectance R, for light normally incident on an interface between two dielectrics is given by

$$R = \left(\frac{n_1 - n_2}{n_1 + n_2} \right)^2 \qquad (1\text{-}17)$$

If light is incident from side 1, there will be a 180° phase shift in the reflected wave for $n_2 > n_1$ but no phase shift if $n_2 < n_1$. This factor is important for antireflection and enhanced reflection coatings on optical components.

For nonnormal angles of incidence, the Fresnel formulas provide the reflectances for parallel and perpendicular polarizations.

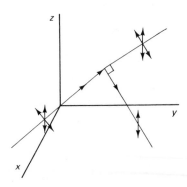

Figure 1-6 Illustration of Brewster's law. Interface is the yz plane and rays are parallel to the xy plane.

$$R_{\parallel} = \left[\frac{\tan (\theta_1 - \theta_2)}{\tan (\theta_2 + \theta_1)} \right]^2$$

$$R_{\perp} = \left[\frac{\sin (\theta_2 - \theta_1)}{\sin (\theta_1 + \theta_2)} \right]^2 \tag{1-18}$$

Note that $R_{\parallel}$ goes to zero for $\theta_2 + \theta_1 = 90°$, the Brewster angle condition.

1-3 MIRRORS AND LENSES

In this section the results derive from the application of the law of reflection to mirrors and Snell's law to thin and thick lenses. Pertinent information concerning spherical aberration is also presented.

The sign convention used here, with regard to lenses and mirrors, follows that of Jenkins and White, 1976. In this convention, light is always assumed to be traveling left to right. All object or image distances measured to the left of a reflecting surface are positive; otherwise they are negative. Object distances to the left and image distances to the right are positive for refracting surfaces; otherwise they are negative. Radii of curvature are positive if measured in the direction of the reflected or refracted light; otherwise they are negative. Lenses or mirrors that converge parallel rays have positive focal lengths and negative focal lengths if they diverge parallel rays. An object height measured above the axis is positive; below the axis it is negative.

Figure 1-7 illustrates the effect of concave (positive) and convex (negative) spherical mirrors on parallel rays.

Parallel rays are reflected through a point called the focal point F for the concave mirror. Point C locates the center of curvature and F lies midway between C and the vertex of the mirror. Hence f, the focal length or distance from F to the vertex, equals $r/2$, where r is the radius of curvature of the mirror. For the convex mirror, rays parallel to the axis are reflected such that they appear to be coming from the focal point F. For the concave mirror, rays parallel to the axis are reflected through the focal point. In both cases,

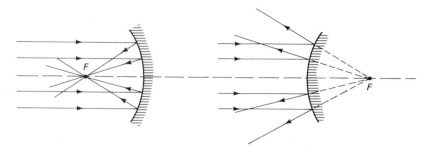

Figure 1-7   The effect of concave and convex mirrors on rays parallel to the axis.

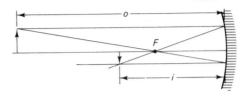

Figure 1-8 Image construction for a spherical mirror.

the rays are reversible. These facts can be used to construct images graphically for a given object. Graphical image construction is depicted in Fig. 1-8. Analytically, object distance o, image distance i, and f are related by

$$\frac{1}{o} + \frac{1}{i} = \frac{1}{f} \tag{1-19}$$

where all distances are measured from the mirror vertex. Also, it is easy to see that magnification m is given by

$$m = \frac{Y_i}{Y_o} = -\frac{i}{o} \tag{1-20}$$

Equations (1-19) and (1-20) apply equally well to spherical mirrors and thin lenses if the sign convention is strictly followed. Actually, not all parallel rays are focused at a common point. This fact leads to what are referred to as aberrations, the most important of which, for our purposes, is spherical aberration. Spherical aberration occurs because rays parallel to the axis are focused closer to the mirror vertex the farther they are from the axis. Also, parallel rays not parallel to the axis suffer an aberration called astigmatism (nothing to do with astigmatism of the eye). Strictly speaking, a spherical mirror is good only for paraxial rays—that is, rays parallel and close to the axis. Parabolic mirrors do not suffer from spherical aberration; all rays parallel to the axis are focused at a common point, but astigmatism is more severe.

Spherical mirrors are little used for precisely focusing purposes except in large telescopes where the focusing is on axis and the spherical aberration can be corrected. Spherical mirrors are used in lasers as resonator mirrors. Here spherical aberration is not a problem because the wavefronts produced are spherical, not plane as we have been tacitly assuming.

Mirrors used for off-axis focusing, such as for high-power lasers, are usually nonspherical or aspheric. This condition is accomplished by using a sector of a parabolic surface as illustrated in Fig. 1-9.

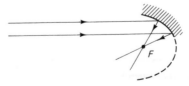

Figure 1-9 Aspheric off-axis focusing mirror.

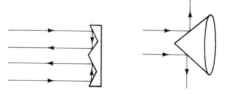

Figure 1–10   Waxicon.   **Figure 1–11**   Axicon.

Many other types of mirrors are used with lasers, such as axicons, so called because they are specifically designed to produce a particular axial distribution of light on reflection (axicons may also be transmissive and are used for spherical aberration correction). A variation, the waxicon (W-shaped axicon), is used to convert the ring mode output of an unstable resonator to a near-Gaussian distribution. Such a mirror is depicted in Fig. 1–10. Another type of mirror focuses the ring mode to a thin ring of finite radius as shown in Fig. 1–11.

Thin lenses can be treated similarly to mirrors except that the analytical results are derived from Snell's law instead of the law of reflection. The thickness of the lens is neglected and two focal points are required. The primary focal point F_p is defined such that light rays coming from it, or headed toward it, are refracted parallel to the axis. The secondary focal point F_s is defined such that rays traveling parallel to the axis are refracted so that they pass through F_s or appear to be coming from it. Figure 1–12 illustrates these comments about focal points for converging and diverging lenses. Note that the refraction is assumed to take place at a single plane referred to as a *principal plane*. The focal lengths are also measured from this plane. If the refractive index of the lens is n' and the refractive indices of the surrounding media are $n = n'' = 1$, then

$$\frac{1}{o} + \frac{1}{i} = \frac{1}{f} = (n' - 1)\left(\frac{1}{r_1} - \frac{1}{r_2}\right) \qquad (1-21)$$

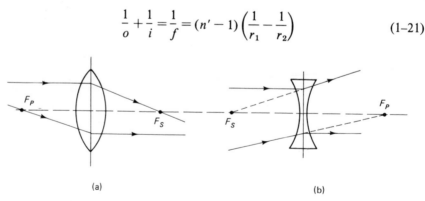

(a)

(b)

Figure 1–12   Primary (F_p) and secondary (F_s) focal point for (a) converging and (b) diverging thin lenses.

where r_1 and r_2 are the radii of curvatures of the first and second surfaces, respectively. The magnification is the same as for the mirrors.

Lenses with spherical surfaces suffer from the same aberrations as mirrors plus several others, including chromatic aberration. Different wavelengths have different refractive indices; therefore different wavelengths are focused at different points. In the context of this book, the monochromatic aberrations, spherical aberration, coma, and astigmatism are the important ones. Coma, like astigmatism, occurs for parallel rays that are not paraxial and causes a lateral smearing of the image. Coma and astigmatism are minimized by keeping the beam as nearly parallel to the axis as possible. Curvature of field and distortion are important in imaging applications.[1] The correction for spherical aberration will be discussed after thick lenses are covered.

In many practical applications, such as focusing of high-power lasers for materials processing, the thin lens assumption is not valid. Not everything previously developed need be discarded, however. In fact, by definition of two principal planes, properly located, the thin lens equation remains valid. The primary focal length is measured from the primary principal plane and the secondary focal length is measured from the secondary principal plane. The principal planes and their relation to the focal lengths and the bending of rays are illustrated in Fig. 1-13.

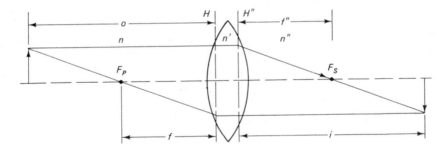

Figure 1-13   Thick lens with principal planes H and H''.

The lens equation becomes

$$\frac{n}{o} + \frac{n''}{i} = \frac{n}{f} = \frac{n''}{f''} \tag{1-22}$$

which is, in fact, the same equation as for a thin lens. It was assumed before that $n = n''$, but it will be instructive to assume $n \neq n''$ for awhile. Note in Fig. 1-13 where the focal lengths are measured from and where the rays appear to bend.

In order to determine the focal lengths of a thick lens, the focal lengths

---

[1] See, for example, *Fundamentals of Optics* by Jenkins and White.

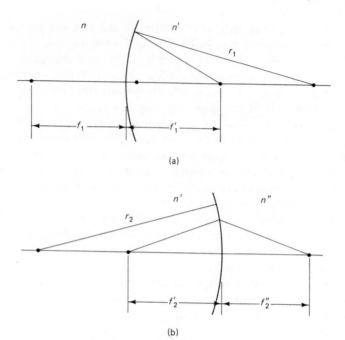

(a)

(b)

Figure 1–14 Refraction at single surfaces.

for refraction at a single surface must be known. The single-surface focal length equations are given as Eq. (1–23) and relate to Fig. 1–14.

$$\frac{n}{f_1} = \frac{n'}{f_1'} = \frac{n' - n}{r_1} \qquad \text{first surface}$$

$$\frac{n_1'}{f_2'} = \frac{n_1''}{f_2''} = \frac{n'' - n'}{r_2} \qquad \text{second surface}$$

(1–23)

Equation (1–24) gives the relationship between the thick lens primary and secondary focal lengths in terms of the single-surface focal lengths and the appropriate refractive indices.

$$\frac{n}{f} = \frac{n''}{f''} = \frac{n'}{f_1'} + \frac{n''}{f_2''} - \frac{dn''}{f_1'f_2'}$$

(1–24)

Equations (1–25) give the distances of the principal planes (H, H'') from the lens vertices, V_1 and V_2, as defined in Fig. 1–13.

$$V_2H = (d - f)\frac{d}{f_2'} \qquad V_2H'' = -f''\frac{d}{f_1'}$$

(1–25)

The procedure for correcting for spherical aberration in spherical lenses is, in effect, to increase the curvature of both refracting surfaces away from

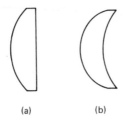

Figure 1–15 (a) Partially corrected plano-convex, (b) optimally corrected meniscus lens for 10.6-μm wavelength.   (a)   (b)

the direction of the oncoming light. An optimum correction exists for all focal lengths and refractive indices, but spherical aberration cannot be eliminated for a single-element lens. Figure 1–15 depicts a partially corrected plano-convex lens and a fully optimized meniscus lens. The light must enter the lens from the most curved side; otherwise spherical aberration is increased. Sherman and Frazier have developed a method for comparing plano-convex and meniscus lens to determine if the superior correction of the meniscus lens warrants its higher cost. The spherical-abberation-limited spot diameter d_{sa} (blur of an ideal focused spot) is given by Eq. (1–26).

$$d_{sa} = k \frac{D^3}{f^2} \tag{1–26}$$

The factor k depends on whether the lens is a plano-convex or meniscus and the refractive index of the lens D, is the diameter of the beam entering the lens containing at least 95% of the beam power. Some values of k are listed in Table 1–1 for $\lambda = 10.6$ μm.

Methods for calculating the diffraction-limited spot size, with which d_{sa} should be compared, are discussed in a later chapter.

1–4 BEAM EXPANDERS

Because of their importance in laser applications, some discussion of beam expanders is presented here.

Lens beam expanders can be made with two positive or one negative and one positive lens. These types of transmissive beam expanders, along with

TABLE 1–1 **Spherical Aberration Factor**
k for Selected Materials at 10.6 μm

Material	Meniscus	Planoconvex
Ge	0.0087	0.0295
ZnSe	0.0187	0.0286
GaAs	0.0114	0.0289

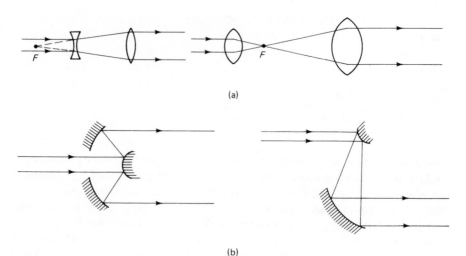

(a)

(b)

Figure 1–16 Beam expanders: (a) transmissive, (b) reflective.

two reflective beam expanders, are depicted in Fig. 1–16. In either case, the lenses have a common focal point. By similar triangles, it is seen from Fig. 1–16 that

$$W_2 = \frac{f_2}{f_1} W_1 \qquad (1\text{–}27)$$

for either lens combination if the absolute values of the focal length of the negative lens is assumed. It is important to note that the diffraction spreading of the beams is reduced in proportion to f_1/f_2 because of the increase in effective aperture.

The nonfocusing beam expander is practical for most applications since it can be made more compact. For very high power applications, it is undesirable to focus the beam in air due to the possibility of breakdown in air or distortion of the beam due to induced refractive index changes. The focusing beam expander is used when beam filtering is required. This process basically involves focusing the beam on a pinhole and filtering (blocking) out the higher-order diffraction fringes. The beam that exits the pinhole contains much less spatial noise and hence has a more uniform irradiance distribution.

1–5 INTERFERENCE AND DIFFRACTION

A brief discussion of interference and diffraction phenomena is presented in this section, particularly as they relate to lasers and laser applications.

When two or more electromagnetic waves having some fixed-phase relationship relative to one another are superimposed, the result is referred to as *interfer-*

ence. The electric field intensities must be added rather than simply adding the powers or irradiances as is done when phase relationships are timewise random, such as with adding the light from two flashlights. If the beam from a laser is split into two beams and then these separate beams are recombined on a screen, as depicted in Fig. 1–17, the phase relation between the beams will vary from point to point on the screen due to slight path length differences. This situation can be expressed mathematically as in Eq. (1–28):

$$E_1 + E_2 = E_{10} \sin (2\pi ft) + E_{20} \sin (2\pi ft + \theta) \tag{1-28}$$

where θ is a phase angle that will vary, depending on location on the screen, and E_{10} and E_{20} are the amplitudes. Since the resultant amplitude is of primary interest, the electric fields can be treated as phasors (vectors in the complex plane with time dependence omitted) with an angle θ between them. Using the law of cosines produces

$$E^2 = E_{10}^2 + E_{20}^2 + 2E_{10}E_{20} \cos \theta \tag{1-29}$$

Since irradiance is proportional to the electric field amplitude squared, Eq. (1–29) can be written

$$I = I_1 + I_2 + 2\sqrt{I_1 I_2} \cos \theta \tag{1-30}$$

It must be remembered that irradiance is a time-averaged quantity and if θ varies randomly with time, the last term in Eq. (1–30) goes to zero, leaving the usual result for addition of noise signals. When θ does not vary randomly with time, an interference pattern is produced on the screen with a visibility V defined by

$$V = \frac{I_{\max} - I_{\min}}{I_{\max} + I_{\min}} \tag{1-31}$$

It is left as an exercise to show that Eq. (1–31) reduces to $2I_1 I_2/(I_1 + I_2)$. Clearly $0 \leq V \leq 1$.

There are numerous ways to produce interference effects with light. Just

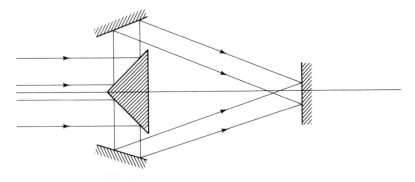

Figure 1–17   Split laser beam recombined on screen.

the propagation of a plane or spherical wave (any shape for that matter) is
the result of continuous interference of an infinity of so-called Huygen's wavelets
that make up the propagating wavefront. Huygen's principle provides a convenient model for the construction of propagating wavefronts given the wavefront
and its direction of propagation at some place and time. According to Huygen's
principle, every point on a wavefront is a source of a spherical wave, with
directionally dependent amplitude, called a *Huygen's wavelet.* Each wavelet expands forward (in the direction of propagation of the original wavefront) at a
rate equal to the velocity of propagation of the original wavefront. The shape
and location of the wavefront after time *t* are determined by constructing wavelets with radius *ct* and drawing a wavefront tangent to all possible wavelets.
Such a construction is depicted in Fig. 1–18.

Figure 1–18 Illustration of Huygen's
principle.

   When such a wavefront is interrupted by an aperture or an edge, some
contributors to the propagating wavefront are removed. The result is a diffraction
pattern such as is commonly produced by a knife edge. This situation is schematically represented in Fig. 1–19. This diffraction pattern can be used for precise
location of a sharp edge.
   Let's look at what happens when a laser beam illuminates two thin rectangular slits in a screen. The experimental setup is depicted in Fig. 1–20. This
particular example is referred to as *Young's double slit interference.* All parallel
rays leaving the slits are focused at a common point in the focal plane of the
lens. The *optical path lengths* traveled by any two parallel rays starting from
a common perpendicular plane are equal. The path length difference between
oblique pairs of rays can be represented as in Fig. 1–20, where a path length
difference of λ is indicated for two parallel rays. These rays interfere constructively in the focal plane, because $\theta = 360°$. If the obliquity is increased until

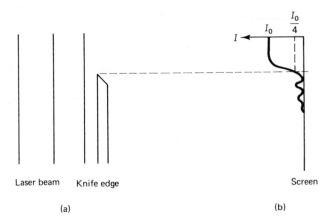

Figure 1–19   Knife edge experiment with laser beam uniformly illuminating the knife edge.

the difference is 2λ or $\theta = 720°$, constructive interference will again be obtained. A general equation can be written

$$m\lambda = d \sin \theta \qquad (1\text{--}32)$$

where $m = 0, 1, 2, \ldots$. Note, also, that $y = f \tan \theta$.

If hundreds or thousands of slits or rulings are provided instead of just two, it is easy to show that Eq. (1–32) still applies and we have what is called a diffraction (interference) grating. The number of rulings (lines) per centimeter is usually given and it equals the reciprocal of d. Because of the multiplicity of lines, nearly total destructive interference occurs for values of θ only slightly different than those given by Eq. (1–32). Hence a diffraction grating can be

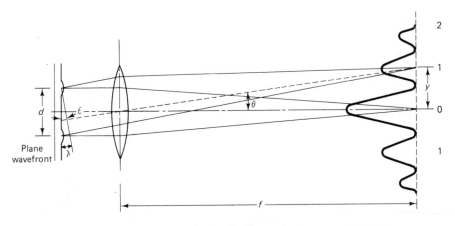

Figure 1–20   Double slit experiment.

used to resolve (separate) closely spaced wavelengths in a beam of light. Diffraction gratings are frequently used as a means of separating laser lines and in wavelength selection, or tuning, in lasers.

Another very important diffraction phenomenon is the effect of an aperture on a beam of light. Figure 1–21 depicts a laser beam incident on a rectangular slit. In this case, it is not as easy to locate the points of constructive interference; so we look for the points of destructive interference, which are of more interest anyhow. The slit is divided into two equal parts of width $D/2$ and a ray is taken from the top of each part. If these rays are parallel and differ in path by $\lambda/2$ or $180°$, then they will interfere destructively as shown. All the rest of the rays, parallel to these two, will also interfere destructively in pairs. To find higher-order minima, simply divide the slit into 4, 6, . . . equal parts and let the path length difference between rays from the tops of adjacent parts be $\lambda/2$. A general equation can be developed (left as an exercise)

$$m\lambda = D \sin \theta \qquad (1\text{--}33)$$

where $m = 1, 2, 3, \ldots$. Note that Eq. (1–33) has the same form as Eq. (1–32) but a different interpretation. The width of the image line produced on the screen is taken to be the distance between first minima, $2y$. For small angles,

$$\lambda = D\theta = \frac{Dy}{f} \qquad (1\text{--}34)$$

An extremely important case of diffraction is that caused by a circular aperture. The derivation for this case is beyond the scope of this book. The result for the first minimum is

$$\sin \theta = \frac{1.22\lambda}{D} \qquad (1\text{--}35)$$

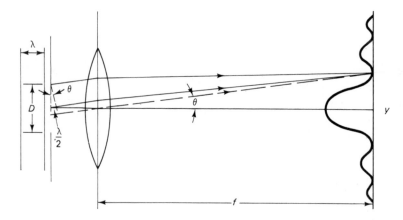

Figure 1–21 Single slit experiment.

Any beam of light made up of *plane waves* truncated by a circular aperture must have a minimum divergence angle given by Eq. (1–35), where D is the truncated diameter. Equation (1–35) is frequently applied to laser beams, letting D be the diameter of the output beam. This is, at best, only a rough estimate because laser beams are not accurately represented by a uniform plane wave.

1-6 COHERENCE

Coherence of electromagnetic waves is a complex subject and no attempt at a rigorous treatment will be made here. The general nature of coherence is relatively simple, however, and a useful working knowledge can be presented fairly easily.

Essentially two waves are coherent if there is some fixed-phase relationship between them. Two pure sinusoidal waves of the same frequency would have perfect coherence. Unfortunately, physical phenomena can, at best, only be approximated by pure sinusoids. In reality, there are two manifestations of coherence, spatial and temporal.

Spatial coherence refers to correlation in phase at the same time but at different points in space. For the experiments on interference and diffraction previously described, a high degree of spatial coherence was necessary for interference or diffraction patterns to occur. Spatial coherence can be produced with ordinary sources in two ways. One is to place the source far enough away from the slits, diffraction grating, and so forth so that each emitting point of the source uniformly illuminates the entire device. The second is to place a small aperture (pinhole) in front of the source, such as a gas discharge lamp. Because of the small emitting area of the source viewed by the diffracting device, substantial spatial coherence is achieved. Lasers emit radiation with nearly perfect spatial coherence across the entire beam.

Temporal coherence refers to correlation in phase at the same point in space at different times. As a wave passes through a given point in space, its phase will undergo random changes. These changes are unpredictable as to when they will occur or as to what the change in phase will be. The origin of these changes is at the source. The average time between changes in phase for an isolated atom is the lifetime of the atom in its excited state, which averages about 10^{-8} s. The average emission time for luminous gases, liquids, or solids, is much shorter due to line-broadening mechanisms. Remember, $\Delta f \tau \simeq 1$ according to the uncertainty relations. For lasers operating in a single mode,[2] the coherence time τ_{coh} can be much longer, approaching that and, in many cases, bettering that of the isolated atom (molecule).

The coherence time is simply the approximate time that it takes a photon

----

[2] Lasers may operate in a variety of axial and transverse electromagnetic (TEM) modes. They are discussed in a later chapter.

to pass a given point in space. Thus the coherence length L_{coh}, which is the approximate length of the photon, is related to the coherence time by

$$L_{coh} = c\tau_{coh} \tag{1-36}$$

The Michelson-Morley experiment exemplifies the significance of temporal coherence. This experiment, which uses the Michelson interferometer, is depicted in Fig. 1–22. Let's assume a laser beam is used as the light source. The beam is split into two equal parts by the beam splitter. Waves 1 and 2 are reflected by the corresponding mirrors and are partially recombined at the beam splitter and observed at the top. In order for interference to be observed over a long period of time, such as required by the eye, a photographic emulsion, or most photodetectors, waves 1 and 2 must overlap. Otherwise there will be no fixed-phase relationship between waves numbered 1 and those numbered 2. Thus

$$2(L_1 - L_2) > L_{coh}$$

there will be no observable interference. This setup has been used to measure coherence length or photon length for decades. Before the advent of the laser, the practical application of the Michelson interferometer was limited to situations where $L_1 \simeq L_2$. Lasers may have coherence lengths easily in excess of 100 m; therefore one leg, the reference leg, can be a few centimeters long whereas the other leg may be nearly 50 m long. Applications of the Michelson interferometer and related interferometric techniques, such as holography, will be discussed in a later chapter.

Some further insight into the concept of coherence and how it can be measured is obtained from the following simplified approach. A beam may be

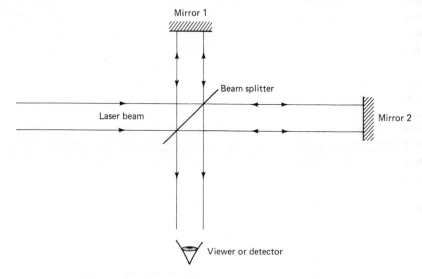

Figure 1–22 Michelson interferometer.

thought of as consisting of a completely coherent part CI and a completely incoherent part $(1 - C)I$. Because there is no correlation between these parts, they can be added like noise signals:

$$I = CI + (1 - C)I$$

where $0 \leq C \leq 1$. If two such beams are superimposed, the coherent parts will add like phasors (coherently):

$$I_{\text{coh}} = (CI) + (CI) + 2\sqrt{CICI} \cos \theta$$

whereas the incoherent parts will add like noise signals:

$$I_{\text{incoh}} = (1 - C)I + (1 - C)I = 2(1 - C)I$$

The total irradiance is obtained by simply adding these last two expressions:

$$I_T = I_{\text{coh}} + I_{\text{incoh}} \tag{1-37}$$

The quantity C could represent the fraction of overlap in a Michelson-Morley experiment or it could represent the degree of spatial coherence in a double slit or other diffraction experiment. It is left as an exercise to show that Eq. (1–37) leads to $V = C$ from the definition of visibility in Eq. (1–30). Thus the fraction of overlap or degree of spatial coherence is easily measured.

1–7 POLARIZATION

The subject of linear polarization has, of necessity, already been dealt with. Because electromagnetic radiation is a vector phenomenon, however, other more general forms of polarization can be obtained from the principle of superposition. Also, some discussion of how different forms of polarized light may be produced is in order.

If you imagine superimposing two electromagnetic waves linearly polarized at right angles to one another but with a 90°-phase difference between them, the result is circular polarization. Figure 1–23 is a series of head-on views of these waves; each point is advanced one-eighth of a wavelength. The tip of the resultant vector is performing uniform circular motion in the plane of $\mathbf{E}_1$ and $\mathbf{E}_2$ and has an amplitude $E_r = E_1 = E_2$. In this view, $\mathbf{E}_2$ lags $\mathbf{E}_1$ by 90° and so left-hand rotation occurs; if the phase relation is reversed, right-hand rotation occurs. The actual path of the tip of the resultant vector is a spiral in the direction of propagation. If $E_1 \neq E_2$, then elliptical polarization occurs with the major axis lying parallel to the larger $\mathbf{E}$ field. If the phase angle is

Figure 1–23 Head-on view of circu-
larly polarized light in $\frac{1}{8}\lambda$ increments,
left-hand rotation.

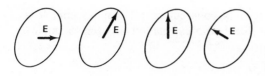

Figure 1–24 Head-on view of elliptically polarized light.

not 90°, elliptical polarization occurs with the major axis lying at some angle between E_1 and E_2. Figure 1–24 depicts elliptically polarized light from a head-on view. The phase angle between E_1 and E_2 is given by

$$\sin \theta = E_b / E_a \qquad (1\text{--}38)$$

regardless of the relative amplitudes of E_1 and E_2. It is left as an exercise to prove Eq. (1–38).

Linearly polarized light is produced in many ways. It occurs in gas lasers when Brewster windows are used to seal the ends of the gas tube or in solid lasers if the ends of the rod are cut at the Brewster angle. Linear polarization also arises when repeated reflections occur from mirrors, for the reflectance is always higher for $E_\perp$ than it is for $E_\parallel$. Lasers with 45° mirrors to fold the beam through a number of tubes have this characteristic. Repeated reflection from a stack of plates oriented at the Brewster angle can linearly polarize an ordinary beam. Polaroid is used extensively for polarizing light and it consists of molecules that preferentially absorb light polarized parallel to their long axis but not their short axis. A sheet of Polaroid is produced by orienting these molecules with all their long axes parallel. The polarizing direction of the material is at 90° to the long axis of the molecules.

Figure 1–25 illustrates an important relationship between polarizing direc-

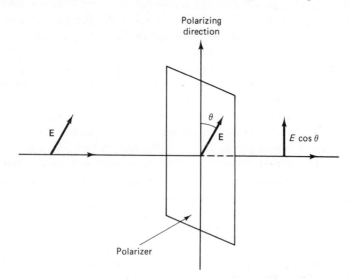

Figure 1–25   Illustration of Malus' law.

tion and irradiance called *Malus' law*. Linearly polarized light is incident on a polarizer at an angle θ relative to the polarizing direction. If the incident amplitude is E_0, the transmitted amplitude is $E_0 \cos \theta$. Therefore the transmitted irradiance is related to the incident irradiance by

$$I = I_0 \cos^2 \theta \qquad (1–39)$$

Several other types of devices produce linearly polarized light from ordinary light more efficiently and with less attenuation than Polaroid, such as Nicole and Glan prisms, which are based on the optical properties of transparent solids. Space does not permit discussion of all these polarizers. Most standard optics books describe them in detail.

It is important to discuss briefly some techniques for producing circular and elliptically polarized light. Many materials exhibit *birefringence*, which means two different refractive indices for the perpendicular components of light passing through the material. Such materials have an optical axis and a unique refractive index occurs for a ray with the electric field vector perpendicular to the optical axis—the normal ray. The refractive index for the other ray, the extraordinary ray, varies from the value for the ordinary ray, depending on the material, up to a maximum or minimum for the electric field vector parallel to the optic axis. One such material is the crystal calcite, which is the basis for the Nicole prism. A ray traveling parallel to the optic axis in calcite travels faster than one perpendicular to it. Because of the differing refractive indices, the two polarized beams can be separated. The property of birefringence is also exhibited by some crystallized polymers and can be induced in a variety of crystals and liquids by application of a strong external electric field. Examples are potassium-dihydrogen-phosphate (KDP) and lithium niobate crystals and nitrobenzene, a liquid. They are referred to as Pockels cells and Kerr cells, respectively. In both cases, high voltages are required, but the optical axis can be switched on and off rapidly, in nanoseconds. Besides, the amount of birefringence (refractive index difference) is a function of applied voltage. Hence the phase difference between the two polarizations, after passing through the medium, can be controlled by the applied voltage.

The difference in the refractive index in nitrobenzene is given by

$$\Delta n = k \lambda E^2 \qquad (1–40)$$

where $k = 2.4 \times 10^{-10}$ cm/V^{-2}. The electrodes are placed parallel to the direction of propagation and the optical axis is perpendicular to the electrodes. The refractive index difference in a Pockels cell is given by

$$\Delta n = pE \qquad (1–41)$$

where $p = 8 \times 10^{-11}$ cm/V^{-1} for deuterated KDP (KD*P) and 3.7×10^{-10} cm/V^{-1} for lithium niobate. The electrodes are perpendicular to the direction of propagation in this type of Pockels cell.

Let's consider a device that alters the phase between two rays by 90°;

such devices are referred to as quarter wave plates. They could be a sheet of birefringent polymeric material, a Pockels cell, or a Kerr cell. Linearly polarized light enters the device such that the direction of polarization makes a 45° angle with the optic axis. Resolving the electric field into components parallel and perpendicular to the optic axis results in two beams traveling at different speeds. This situation is depicted in Fig. 1–26. Normally a fast axis and a slow axis are specified in the material. The beam with polarization parallel to the fast axis advances just one quarter wavelength, 90°, relative to the other beam on passing through the material. Hence the beam exiting the quarter wave plate is circularly polarized. If E is not 45° to the optic axis, then elliptically polarized light is produced. If the thickness corresponds to some phase angle difference other than 90°, elliptical polarization occurs. A phase change of $\lambda/2$ or 180° simply rotates the direction of polarization by 90°. These are important factors in many laser applications, such as rapid Q-switching of lasers and conversion of a linearly polarized laser beam to circularly polarized to improve cutting ability.

An interesting consideration is that of the superposition of two counter-rotating, circularly polarized light beams. If the beams have equal amplitudes, the result will be linearly polarized light with twice the amplitude of one of them. Figure 1–27 will aid in explaining this situation. As the electric field vectors rotate, their sum remains parallel to the bisector of the angle between them. If the phase of one of the beams is changed, however, say by moving a mirror from which it is reflected, the orientation of the resultant **E** vector will be changed. A $\lambda/8$ motion of the mirror causes a $\lambda/4$, 90° phase change, which produces a 45° rotation of the resultant **E** vector. Not only can this be used to determine how far the mirror has moved, but the direction of rotation of the resultant **E** vector indicates which way the mirror has moved.

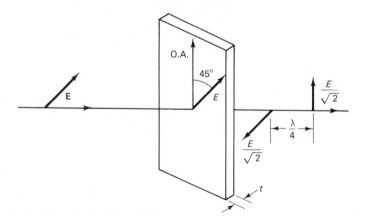

Figure 1–26 Quarter waveplate. O.A. is the optical axis and t is appropriate thickness to provide 90° phase shift ($\lambda/4$) between the orthogonal components.

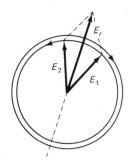

Figure 1–27 Superimposed counter-rotating, circularly polarized beams. (Drawn slightly different amplitudes for clarity.) E_r is the resultant and is linearly polarized with amplitude equal to $2E_1$ for $E_1 = E_2$.

A final phenomenon of particular interest when it is essential that light reflected from optical components not be fed back into the laser is the Faraday effect. Some materials, such as lead glass, become *optically active* when subjected to a strong magnetic field parallel to the direction of propagation. Optical activity means that linearly polarized light entering this material will have its polarization direction rotated. The amount of rotation is given by

$$\phi = VBl \qquad (1\text{–}42)$$

where B is the magnetic field intensity, l is the length of material, and V, the Verdet constant, has a value of 110 (tesla·meter)$^{-1}$ for dense flint glass.

If linearly polarized light is passed through a Faraday rotator that rotates the polarization direction 45°, any light that is reflected back through the rotator will be rotated an additional 45°, for a total of 90°. This light, traveling in the reverse direction back toward the laser, is easily attenuated by a polarizer oriented to transmit the original light beam.

1–8 FIBER OPTICS

The subject of fiber optics is included here because of the growing use of fiber optics in optical instrumentation, communication, and laser beam delivery for surgical and other important applications. Optical fibers may be either glass or plastic. When low losses are important, glass fibers are superior, having losses of less than 1 dB/km of fiber length available.[3]

The operation of simple optical fibers is based on the phenomenon of total internal reflection. If a fiber core is clad with a material of lower refractive index than the core, there will always be a critical angle such that rays striking the interface of core and cladding at an angle greater than this angle will be totally reflected. Thus the fiber becomes a light guide or dielectric waveguide. This phenomenon is illustrated in Fig. 1–28.

In Fig 1–28 a ray is shown entering the face of the fiber at the maximum

---

[3] The loss in decibels is defined as $10 \log_{10} P_\text{in}/P_\text{out}$, where P_in and P_out are the input and output optical power, respectively.

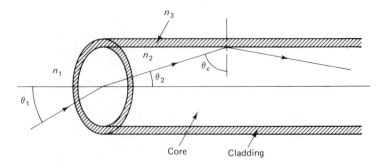

Figure 1–28 Optical fiber showing entering ray striking cladding at critical angle.

possible angle that will result in total internal reflection of the ray inside the fiber. Applying Snell's law at the fiber face and assuming the ray strikes the core-cladding interface at the critical angle, we can show that

$$\text{NA} = n_1 \sin \theta_1 = (n_2^2 - n_3^2)^{1/2} \tag{1–43}$$

Here n_1, n_2, and n_3 are the refractive indices of the external medium, core, and cladding, respectively, θ is the half-angle of the entrance cone, and NA stands for numerical aperture.

The cladding in glass fibers is also glass but doped differently than the core to give it a slightly lower refractive index. For example, if the core and cladding indices are 1.55 and 1.50, respectively, for a fiber in air, $\text{NA} = 0.39$ and $\theta_1 = 23°$.

The type of fiber just discussed is a step index fiber. Another type commonly used is the graded index fiber. Here the refractive index decreases radially from the center in a parabolic fashion. Rays launched at small angles to the axis of the fiber continuously change direction toward the axis without ever striking the cladding. The path is a zigzag combination of straight-line segments between reflections in a step index fiber whereas the path in a graded index fiber is essentially sinusoidal.

Losses in glass fibers are primarily due to scattering by impurities and defects. Losses in plastic fibers are chiefly due to absorption.

Short delay times in optical communications are desirable to minimize pulse spreading and the resultant degradation of information. There are two causes of delay in optical fibers, referred to as modal dispersion and wavelength dispersion. Modal delay is caused by the different distances traveled by different rays within the fiber. The maximum value for this type of delay in a step index fiber can be deduced by considering a ray traveling parallel to the axis and a ray reflecting from the cladding at the critical angle. The time delay per unit of fiber length is then given by

$$\Delta t = \frac{n_2}{c} \left(\frac{1}{\sin \theta_c} - 1 \right) = \frac{n_2}{c} \left(\frac{n_2}{n_3} - 1 \right) \tag{1–44}$$

This amounts to 0.17 ns/m for the fiber example discussed. Actual delay times are less because the losses are higher for the rays traveling greater distances. Optical fibers are waveguides and, as such, not every imaginable geometrical path for a ray is an allowable mode. In fact, if the fiber is made sufficiently thin, typically a few micrometers in diameter, only one allowed mode exists. Such single-mode fibers effectively eliminate modal delay, and picosecond delay times are possible. Graded index fibers minimize modal dispersion because the farther a ray wanders from the axis, the more time it spends in a region of lower refractive index (higher speed). Thus the time delay between axial and nonaxial rays is reduced.

Wavelength dispersion, or delay due to the variation of refractive index with wavelength, is minimized by proper material design to reduce this variation and by the use of narrow bandwidth light emitters, chiefly diode lasers.

It must be recognized that losses also occur due to reflection at entrance and exit surfaces and coupling of light from emitters into fibers and the coupling of light out of fibers to detectors. Highly efficient techniques and devices have been developed, however, for coupling light in and out of fibers and for coupling light from one fiber to one or several other fibers.

1–9 ACOUSTO-OPTIC MODULATORS

Acousto-optic modulators are used for the production of short, high-power pulses from lasers, beam deflection, amplitude modulation, and frequency modulation. The variety of laser applications makes a discussion of such devices pertinent.

An acousto-optic modulator is a device that uses a transparent solid block of material, such as fused quartz, to which is attached a piezoelectric transducer. Acoustic waves are produced in the quartz by applying a high-frequency voltage, say 40 MHz with 0.5 V amplitude, to the transducer. The vibration of the piezoelectric transducer sends acoustic waves through the quartz block. This process is depicted in Fig. 1–29. A laser beam is passed through the block at a small angle relative to the acoustic wavefronts. Light reflected from the wavefronts can interfere constructively only if Eq. (1–45) is satisfied,

$$m\lambda_0 = 2\lambda_a \sin \theta, \qquad m = 1, 2, 3, \ldots \qquad (1\text{--}45)$$

where λ_0 is the optical wavelength and λ_a is the acoustic wavelength. With care, only the first-order, $m = 1$, deflection occurs along with the original beam. Equation (1–45) may be recognized as similar to Bragg's law for x-ray or electron diffraction from crystals, where λ_a would be the interplanar spacing.

Amplitude of the diffracted beam can be varied from zero to a maximum by varying the voltage applied to the piezoelectric transducer.

The frequency of the diffracted beam is shifted as a result of the Doppler

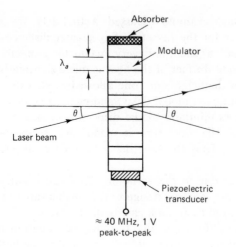

Figure 1-29 Acousto-optical modulator.

effect. According to relativity theory, when an observer and a light source have a relative velocity v, the observed frequency is given by

$$f = f_0 \frac{1 + (v/c)}{\sqrt{1 - (v^2/c^2)}} \qquad (1\text{-}46)$$

where f_0 is the frequency if observer and source are at rest relative to each other. In this case, $v \ll c$ and so Eq. (1-46) may be written

$$\frac{\Delta f}{f_0} = \frac{v}{c} \qquad (1\text{-}47)$$

The velocity is positive for relative motion toward each other and negative for motion away from each other. If the light is reflected from a moving mirror, the Doppler shift is doubled. In the acousto-optic modulator

$$\frac{\Delta f}{f} = \frac{2 v_a \sin \theta}{c} \qquad (1\text{-}48)$$

because the component of acoustic velocity in the beam direction is $v_a \sin \theta$. Combining Eqs. (1-48) and (1-45) for $m = 1$ gives

$$\Delta f = \pm f_s \qquad (1\text{-}49)$$

The plus sign is taken if the acoustic wave velocity component is opposite the beam propagation direction and the minus sign is taken if the acoustic velocity has a component in the direction of beam propagation.

Acousto-optic modulators can be used for deflection (usually less than 0.5°) with over 1000 resolvable spots, amplitude modulation, and frequency modulation.

PROBLEMS

1-1. Show that

$$E = E_0 \sin (\omega t - kx) \quad \text{and} \quad E = E_0 e^{j(\omega t - kx)} \quad (j = \sqrt{-1})$$

are solutions of Eq. (1-1) and determine the relationship between c, ω, and k.

1-2. A laser beam of 2.5 kW average power is focused to a spot of 50 μm diameter in a medium whose refractive index is 1.5. Determine the irradiance and the electric field amplitude.

1-3. Copper has an electrical conductivity of 6×10^7 (ohm·m)$^{-1}$. Calculate the magnitude of copper's intrinsic impedance at the frequency of a HeNe laser ($\lambda = 0.6328$ μm).

1-4. Calculate the skin depth for the data given in Problem 1-3.

1-5. Calculate the theoretical reflectance for highly polished copper for Nd-YAG laser radiation ($\lambda = 1.06$ μm).

1-6. (a) Calculate the momentum of a photon with a wavelength of 10 μm (CO_2 laser).

(b) Calculate the pressure exerted by a 1-terrawatt beam focused to a spot of 150 μm diameter.

1-7. Calculate the Brewster angle and the critical angle for light traveling from a medium of 1.5 refractive index (glass) into a medium of 1.33 refractive index (water).

1-8. Brewster windows are to be attached to the ends of a gas laser tube. The windows have a refractive index of 1.5 and are to be oriented so that any light reflected from them is totally polarized perpendicular to the plane of incidence (hence the name). At what angle relative to the tube axis should these windows be oriented?

1-9. Calculate the pertinent reflectances for light incident normally and at 35° to the normal for light traveling in air incident on a medium of 1.4 refractive index.

1-10. An object is placed on the axis 50 cm from the vertex of a concave spherical mirror that has a radius of curvature of 30 cm. Locate the image and determine the magnification. Illustrate with a ray diagram.

1-11. Repeat Problem 1-10 for a convex mirror of the same radius of curvature.

1-12. By careful drawing of a ray diagram, demonstrate spherical aberration for a concave spherical mirror.

1-13. An object is placed on axis 100 cm from a 20-cm focal length positive thin lens. Locate the image and calculate the magnification. Illustrate with a ray diagram.

1-14. Repeat Problem 1-13 for a negative lens of -20-cm focal length.

1-15. A thick meniscus lens of ZnSe used for focusing high-power CO_2 laser beams has a refractive index of 2.4, thickness of 0.92 cm, and radii of curvature of $r_1 = 11.2$ cm and $r_2 = 30$ cm. Locate the principal planes and determine the focal length of the lens (assume that the lens is to be used in air).

1-16. Calculate the spherical aberration limited spot size of a CO_2 laser beam ($\lambda = 10.6$ μm) of 2-cm diameter (95% of the beam power falls in this area) focused by a 12.5-cm focal length GaAs lens for both meniscus and plano-convex lenses (see Table 1–1).

1–17. Design two types of beam expanders, using lenses that will expand a 2.5-cm-diameter beam to 10 cm. Do not use a lens with a focal length whose absolute value is less than 10 cm.

1–18. Two coherent laser beams interfere with one another on a screen. The irradiances are 10 mW/cm² and 15 mW/cm². Calculate I_{max}, I_{min}, and V. Calculate I at a phase angle difference of 30°.

1–19. Assuming that the diffraction spreading (divergence) of a HeNe laser beam ($\lambda = 0.6328 \ \mu$m) can be approximated by assuming it is equivalent to the diffraction caused by a plane monochromatic wave incident on a circular aperture of diameter equal to the laser beam diameter, calculate the divergence for a 2-mm-diameter beam.

1–20. The frequency bandwidth of a HeNe laser (0.6328 μm) is 100 MHz. Estimate its coherence time and coherence length.

1–21. A crossed polarizer is placed in front of a plane-polarized light beam. A second polarizer is placed between the first and the light source such that its polarizing axis makes an angle of 30° with the E field of the incident light. Assuming 10 mW of light power incident on the second polarizer, how much power would be detected after the light has passed through both polarizers?

1–22. Calculate the refractive index difference per centimeter in lithium niobate ($LiNbO_3$) for an applied field of 1000 V/cm. How long must a longitudinal Pockels cell be to cause a phase change of 90° between the fast and slow rays for ruby laser light ($\lambda = 0.6943 \ \mu$m)?

1–23. Calculate the numerical aperture and half-angle of the entrance cone for light entering a step index optical fiber with core refractive index of 1.55 and cladding index of 1.45. Assume that the fiber is in air.

1–24. Calculate the modal dispersion time delay per kilometer of length for the fiber in Problem 1–23.

1–25. What angle should a HeNe laser beam make relative to the normal, to the surface of an acousto-optical modulator operated at 40 MHz with an acoustical wavelength of 60 μm? What is the shift in frequency of the optical wave, assuming the optical wave has a propagation component in the same direction as the acoustical wave?

REFERENCES

JENKINS, J. A., AND W. E. WHITE, *Fundamentals of Optics* (4th ed.). New York: McGraw-Hill Book Co., 1976.

BIBLIOGRAPHY

1–1. Jackson, J. D., *Classical Electrodynamics.* New York: John Wiley & Sons, Inc., 1962.

1–2. Ziock, K., *Basic Quantum Mechanics.* New York: John Wiley & Sons, Inc., 1969.

1–3. Young, M., *Optics and Lasers.* Berlin: Springer-Verlag, 1977.

1–4. Nussbaum, A., and R. A. Phillips, *Contemporary Optics for Scientists and Engineers.* Englewood Cliffs, NJ: Prentice-Hall, Inc., 1976.

1–5. Williams, C. S., and C. A. Becklund, *Optics: A Short Course for Engineers and Scientists.* New York: John Wiley & Sons, Inc., 1972.

1–6. Smith, W. J., *Modern Optical Engineering.* New York: McGraw-Hill Book Co., 1966.

1–7. Lacy, E. A., *Fiber Optics.* Englewood Cliffs, NJ: Prentice-Hall, Inc., 1982.

1–8. Mansell, D. N., and T. T. Saito, "Design and Fabrication of a Nonlinear Waxicon," *Optical Engineering,* **16,** No. 4., July–August 1977, 355–359.

1–9. Arnold, J. B., R. E. Sladky, et al., "Machining Nonconventional-Shaped Optics," *Optical Engineering,* **16,** No. 4, July–August 1977.

chapter 2

Solid-State Concepts

The purpose of this chapter is to provide the reader who does not already have a strong enough solid-state background with an overview of the concepts required for an understanding of semiconductor optical devices. A short review of some atomic physics concepts is presented and then the concept of energy bands in crystals is discussed and related to the *pn* junction and transistors. Some simple small-signal amplifier circuits are described.

2-1 ATOMIC STRUCTURE

The Bohr theory of the hydrogen atom (actually any one-electron system), particularly its successor, the Bohr-Sommerfeld theory, was eminently successful in providing a model of electronic structure of the atom. Although this theory is not valid for systems with more than one electron, the scheme of orbitals and suborbitals devised for hydrogen serves conceptually as a starting point for building up the electronic structure of all elements of the periodic table. The quantum mechanics later developed by Schrödinger and Heisenberg proved to be the correct nonrelativistic theory of the atom but, except for a difference in the interpretation of orbital angular momenta, leads to the same results for hydrogen as the Bohr-Sommerfeld model. It is the results of the work of Schrödinger and Heisenberg that are discussed here. The quantum mechanical analysis of the hydrogen atom leads to the conclusion that the electron associated with the nucleus can occupy various energy levels for finite periods of time without emitting radiation. When the electron makes a transition from a higher-energy level E_2 to a lower-energy level E_1, the energy given off as radiation is

$$E_2 - E_1 = hf$$

The various states that the electron can occupy are described by a set of four quantum numbers. Three deal with the three dimensions, or degrees of freedom, required to describe the electron's motion around the nucleus and the fourth relates to the intrinsic angular momentum of the electron. The intrinsic angular momentum of the electron is a relativistic phenomenon and does not appear in the Schrödinger or Heisenberg formulations but does in the later relativistic development due to Dirac. These four quantum numbers are designated by n, l, m_l, and m_s. As implied, the first three are a natural result of the solution of the Schrödinger wave equation for the hydrogen atom and the fourth, in general, is simply tacked on to account for the relativistic nature of electrons. The principal quantum number n determines the shells, in chemical or spectroscopic terminology, and has integer values from one to infinity. Symbolically the numbers $n = 1, 2, 3, 4, \ldots$ are represented by $K, L, M, N, \ldots$ etc. The principal quantum number physically relates to the average distance of the electron from the nucleus and its energy.

The quantum number l can take on values $l = 0, 1, 2, \ldots n - 1$ and determines the orbital angular momentum of the electron

$$L = \sqrt{l(l + 1)}\,\hbar$$

($\hbar = h/2\pi$) and its energy to some extent. Note that the angular momentum is zero for $l = 0$, suggesting that the electron travels in a straight line back and forth through the nucleus (in the Bohr-Sommerfeld theory $l = 1, 2, 3, \ldots n$). It presents no conceptual difficulty, for according to DeBroglie, both the electron and the nucleus behave like waves and can therefore pass through one another, like electromagnetic waves or water waves, without necessarily interacting.

The third quantum number, m_l, is the magnetic quantum number and has values of $m_l = -l, -(l - 1) \ldots 0 \ldots (l - 1), l$. Physically this quantum number gives the projection of the orbital angular momentum onto an applied (internal or external) magnetic field. L can have projections on a magnetic field of $-l\hbar \ldots 0 \ldots l\hbar$. The reason L tends to line up with a magnetic field is because of the magnetic moment associated with the orbital angular momentum of the electron. Unlike a classical magnet, the magnetic moment associated with the electronic orbital motion cannot align itself perfectly with the magnetic field but can take only certain orientations as prescribed by the quantum rules.

The final quantum number, m_s, sometimes called the spin quantum number (even though there is no such classical analog), has values of $\pm\frac{1}{2}$ and physically gives the orientation of the electron's intrinsic angular momentum relative to an applied magnetic field. The allowed projections are $\pm\frac{1}{2}\hbar$. The intrinsic angular momentum of the electron is

$$\left[\frac{1}{2}\left(\frac{1}{2}-1\right)\right]^{1/2}\hbar = \left(3/4\right)^{1/2}\hbar$$

It is easily seen then that the K shell ($n = 1$) has $l = 0$, $m_l = 0$, and $m_s = \pm\frac{1}{2}$. At this point we invoke the Pauli exclusion principle, which says that *no two electrons can occupy the same quantum state*. In other words, no two electrons can have the same set of quantum numbers if they are in the same system. A system might be a single atom, molecule, or an entire crystal. Therefore the K shell can have, at most, two electrons in it. Here we presume that this quantum number scheme, derived for hydrogen, will apply to atoms with many electrons, although we do not expect the energy levels to be the same.

Following this same reasoning, the L shell has $l = 0$, 1, $m_l = -1$, 0, 1, and $m_s = \pm\frac{1}{2}$ for each distinct set of l and m_l, providing for a total of eight electrons. The maximum number of electrons in any shell is $2n^2$.

In spectroscopic terms, the various values of l represent subshells within each shell and are denoted, for historical reasons, by s, p, d, f, g, . . . for $l = 0, 1, 2, 3, 4, 5, ...$ Thus the electronic structure of sodium can be written

$$1s^2 2s^2 2p^6 3s^1$$

where the numbers represent the principal quantum numbers or shells and the letters represent the l values or subshells. The superscripts are the number of electrons in each subshell. Sodium has an atomic number of 11; so 11 electrons should be associated with a neutral sodium atom.

It is the physical nature of atoms, like all natural systems, that they tend toward their lowest energy configuration which involves achieving full subshells. Sodium can do so either by gaining another electron or by giving one up, as it does in NaCl and in its metallic form. In NaCl the chlorine atom picks up the electron to close its $3p$ subshell, thus forming an ionic bond with sodium. Materials with ionic bonds are insulators.

In such metals as sodium or copper the valence electrons are mutually shared by atoms in a large region of the material. It is this mutual sharing of electrons that provides the binding forces, the so-called metallic bond.

It is easy to see why sodium and copper are metals: both have unfilled s subshells. It is not so obvious why magnesium, which has closed subshells, is a metal or why silicon, which at a casual glance does not appear to have filled subshells, is an insulator at low temperatures and conducts electricity at room temperature. In order to explain metals like magnesium and semiconductors like silicon, the energy band structure of crystals is introduced.

2-2 ENERGY BANDS

To develop the notion of energy bands qualitatively, we will return to sodium as an example. Imagine 10^{22} sodium atoms so widely separated that they do not interact. There are 10^{22} electrons occupying 2×10^{22} separate $3s$ states. Now imagine gradually bringing these 10^{22} atoms together to form a single crystal with a volume of about 1 cm^{-3}. As they approach each other within a few atomic radii, the outer energy levels of atoms begin to overlap and become perturbed (this does not mean mad at each other). The perturbation causes the energy levels to shift, spreading out into a band of 2×10^{22} different, but closely spaced, energy levels. Figure 2-1 is a schematic representation of this thought experiment. Here the atoms start out at $r = \infty$ and end up at $r = r_0$, the equilibrium spacing of the atoms. The original 2×10^{22} $3s$ levels spread out into a band of 2×10^{22} separate levels shared by the entire assemblage of atoms. According to the Pauli exclusion principle, this band will be half filled by electrons. This means that there are 1×10^{22} unoccupied energy states into which electrons can move to take part in electrical or thermal conduction. Hence sodium is an electrical conductor (metal in this case) and has an electronic contribution to its thermal conductivity.

Note that the p levels in Fig. 2-1 also spread out into an energy band that overlaps the $3s$ band. Thus sodium would have been a conductor even if the $3s$ band had been filled, which explains why the elements of the second column of the periodic table, like magnesium, are metals.

Materials like diamond, germanium, silicon, and gallium arsenide are more difficult to explain due to the nature of the covalent bonding in these materials.

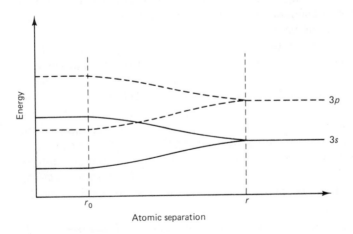

Figure 2-1 Energy levels of Na as a function of atomic separation, r_0 is the equilibrium atomic separation.

Covalent bonding is the mutual sharing of electrons between atoms. Consider carbon, which has atomic number six. We might expect its electronic structure to be $1s^2 2s^2 2p^2$. It turns out that for purely quantum mechanical reasons that are beyond the scope of this book, the $2s$ and $2p$ levels hybridize into two separate sets of four sp^3 levels.[1] When the crystal is formed, these separate sets of levels form nonoverlapping bands with equal numbers of states in each one. Because the total number of hybrid sp^3 states per atom is eight, there will be four states per atom in each of the bands. If the energy gap between the bands is large, the four electrons per atom will go into the lower energy band. In diamond the energy bands are separated by a gap of about 6 eV.[2] Since the average thermal energy at room temperature is about 0.01 eV, very few electrons are able to jump the gap to take part in electrical or thermal conduction. Therefore, diamond is an excellent electrical insulator. (It happens to be an excellent thermal conductor, too, but obviously for a different reason.) When the gap between the separate sp^3 bands is not so great, the result is a semiconductor, such as silicon or germanium. Essentially the compound semiconductors like gallium arsenide are semiconductors for the same reason.

The distinction between metals, semiconductors, and insulators can be summarized in a simplified energy band scheme as given in Fig. 2–2. A partially filled band is indicative of a metal; a material with a relatively small energy gap will be a semiconductor; a material with a relatively large energy gap is an insulator. Note, however, that at a high enough temperature, assuming they do not melt first, all insulators become semiconductors.

Physically the energy gap in a semiconductor or insulator represents the energy required to break a covalent bond. When sufficient thermal energy is accumulated to break a bond, the electron moves to the conduction band. There is now an unoccupied level in the valence band. This unoccupied state, or incomplete covalent bond, has a positive charge associated with it and is easily dissociated from the original site and becomes free to take part in conduction, much the same as the electron. This positively charged quasi particle is called a hole. So when a bond is broken, an electron-hole pair (EHP) is produced. Such EHPs can also be produced by absorption of a photon of sufficient energy, which is the basis for solar cells and a variety of photodetectors.

Up to this point the nature of pure semiconductors has been discussed. Such semiconductors are called intrinsic semiconductors because the properties are intrinsic to the material. Most interesting properties of semiconductors arise from the addition of impurities. The latter semiconductors are referred to as extrinsic. To illustrate the addition of impurities to semiconductors, silicon (Si) will be used as an example. Silicon forms the diamond structure by sharing

---

[1] These hybrid states are denoted by sp^3 because one of the s electrons is promoted to a p level and four distinct orbitals are formed.

[2] An eV is the kinetic energy supplied one electronic charge when accelerated through 1 volt. 1 eV = 1.6×10^{-19} J.

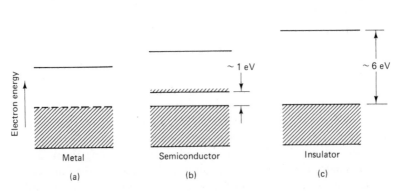

Figure 2-2 Simplified energy band scheme for metals, semiconductors, and insulators.

its four valence electrons with four surrounding atoms in a tetrahedral coordination. Some Si atoms in the crystal can be substituted for by atoms of similar size, such as boron (B) or phosphorus (P), which have valences of 3 and 5, respectively. When P is substituted, the one electron left over is easily dissociated from the parent atom to become a free electron in the conduction band. The P atom becomes a negative fixed ion and no hole is produced in the process. Since the number of intrinsic EHPs at room temperature is about 1.5×10^{10} cm^{-3}, the number of atoms about 5×10^{22} cm^{-3}, even a light impurity concentration of one part per million (ppm) or 5×10^{16} cm^{-3} means that there will be about a million more extrinsic electrons than intrinsic EHPs. Thus the predominant charge carrier will be electrons and this extrinsic material is labeled n type for negative-charge carriers. Actually, the number of holes diminishes markedly because of the "law of mass action," which requires that

$$np = n_i^2 \qquad (2-1)$$

where n is the number of electrons per unit volume, p the number of holes per unit volume, and n_i the number of electrons or holes per unit volume in intrinsic material. Thus for the example given,

$$p = \frac{(1.5 \times 10^{10})^2}{0.5 \times 10^{16}} = 4.5 \times 10^4$$

or 12 orders of magnitude less than n. The holes can be ignored.

If Si is doped with B, then a p-type (positive-charge carriers) material is produced by the same reasoning as for the n-type material. The B atom has a valence of 3 and cannot complete the four bonds with its neighbors. The incomplete bond is easily dissociated from the parent B atom, making it a negative ion and producing a hole in the valence band capable of electrical and thermal conduction.

The properties of p-type and n-type semiconductors by themselves are of limited value. However, when combined in pairs—that is, pn junctions—a myriad of electronic and optical devices becomes possible.

2–3 pn JUNCTIONS

Metallurgical junctions between p- and n-type materials are produced by diffusion, epitaxial growth, and ion implantation. In all cases, the junction is made in a single crystal. It is instructive, however, to imagine starting with two separate crystals, one n type, the other p type, and suddenly joining them to form one continuous single crystal with the metallurgical junction at the interface. Figure 2–3 depicts the various steps in this thought experiment. At the instant the materials are joined there is an infinite concentration gradient across the junction for both holes and electrons. Consequently, holes and electrons *diffuse* in opposite directions, producing a large current until sufficient fixed ionic charges are uncovered in the newly created depletion region (devoid of free charge) to produce a counter-electric field that causes an opposing drift current. The following current balance holds true at equilibrium.

$$I_{\text{dif elec.}} = I_{\text{drift elec}} \tag{2–2}$$

$$I_{\text{dif holes}} = I_{\text{drift holes}}$$

Equations (2–2) state that at equilibrium the diffusion current of electrons equals the drift current of electrons and that the same is true for holes. In terms of the energy band scheme, this situation can be represented as in Fig. 2–4, where only the lower conduction band and upper valence band edges are shown. E_{F}

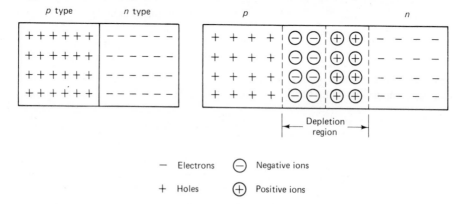

Figure 2–3 pn junction thought experiment. When the junction is formed, holes and electrons diffuse across the junction, uncovering bound ionic charge in the depletion region. The resulting built-in electric field creates drift currents that balance the diffusion currents.

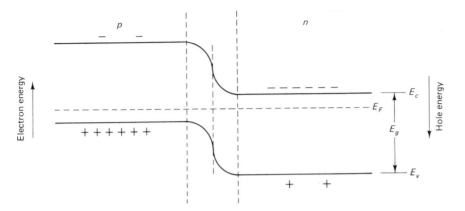

Figure 2-4 Energy band edges for a *pn* junction at equilibrium.

is called the Fermi energy and for our purposes is a reference energy that lies about in the middle of the energy gap for intrinsic materials and is shifted toward the valence band for *p*-type and toward the conduction band for *n*-type materials. E_F is constant throughout a semiconductor at equilibrium; thus it aids in drawing energy level diagrams.

The electric field built into the depletion region as a result of the uncovered ions produces a potential energy barrier for electrons and holes, preventing substantial diffusion of either charge carriers from their majority areas into the minority regions. The *pn*-junction device being considered here is a diode and can be either forward or reverse biased by placing voltages across it as depicted in Fig. 2-5. When the diode is forward biased, the Fermi level is shifted upward on the *p* side, downward on the *n* side, thus lowering the potential energy barrier and allowing a large increase in the *diffusion* of holes and electrons across the junction. A resistor is placed in this circuit to limit current. The drift current is decreased, but this factor has relatively little effect because it

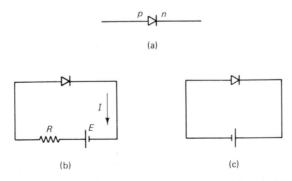

Figure 2-5 (a) Diode symbol. (b) Forward-biased diode. (c) Reverse-biased diode.

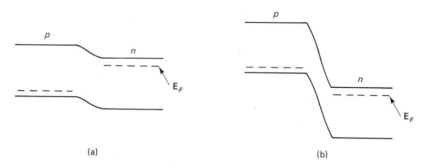

Figure 2–6 Energy band edges for (a) forward-biased and (b) reverse-biased diodes.

is a current of minority carriers and there are few to begin with. When the diode is reverse biased, the shifts in the Fermi level are reversed, thereby increasing the potential energy barrier. Although there is a small increase in the drift current, it is relatively unimportant because it is a minority current and cannot be increased to any very large absolute value. In reverse bias the diode acts like an extremely high resistance. The energy level picture of forward and reverse biasing is given in Fig. 2–6. The diode current versus diode voltage curve (characteristic curve) is sketched in Fig. 2–7 for Si. Note that the current scales are different for forward and reverse voltages.

A Si diode is said to "turn on" at about 0.55 to 0.7 V forward bias; 0.6 V will be used here for convenience. Large increases in current occur for voltages slightly larger than the "turn-on" voltage. The reverse current is relatively constant and usually less than 1 nA until reverse breakdown occurs. This phenomenon will be discussed subsequently. The characteristic curve, exclusive of the breakdown region, obeys the following diode equation closely:

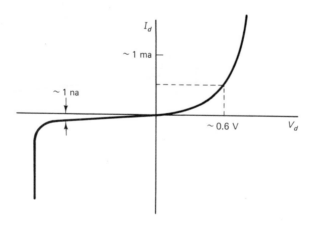

Figure 2–7 Diode characteristic curve for Si, showing reverse breakdown.

$$I = I_0 \left(\exp \frac{eV}{kT} - 1 \right) \tag{2-3}$$

where I_0 is the reverse saturation current, e the electronic charge, V the voltage across the diode, k Boltzmann's constant, and T absolute temperature. A convenient number to remember is that $kT = 0.026$ eV at $T = 300°$K.

The reverse breakdown is caused by two phenomena. At low voltages it is caused in extremely heavily doped diodes by quantum mechanical tunneling of electrons through a very thin depletion layer. At higher voltages in less heavily doped diodes the cause is an avalanche effect in which electrons passing through the depletion layer acquire sufficient energy, because of the large electric field produced by the reverse bias, to ionize neutral atoms in the depletion layer. At a sufficiently high voltage the result can be an avalanche or multiplying effect. Neither type of breakdown is inherently harmful. Damage occurs if the heat generated by a large breakdown current is not dissipated rapidly enough to prevent melting of the diode. Reverse breakdown is used for voltage control and as the basis of avalanche photodiodes.

Different types of semiconductor diodes have different turn-on voltages and reverse saturation currents. Some values of these quantities are listed in Table 2–1.

TABLE 2–1   Reverse Saturation Currents and "Turn-on" Voltages for Some Important Semiconductor Diodes

	Reverse Saturation Current	Turn-on Voltage (Volts)
Si	<1 nA	0.6
Ge	<1 μA	0.3
GaAs	–	1.2
$GaAs_{.6}P_{.4}$	–	1.8
GaP	–	2.2

In determining the resistor size to use in a circuit in order to limit the forward current through a diode, it is customary to assume a voltage drop across the diode equal to its "turn-on" voltage. This situation is illustrated in Fig. 2–8 for a Si diode. The resistance in Fig. 2–8 is given by

$$R = \frac{12 \text{ V} - 0.6 \text{ V}}{0.050 \text{ } A} = 288 \text{ } \Omega$$

In order to limit the current during reverse breakdown, we would do the same as in Fig. 2–8 except that the precise value of the reverse breakdown voltage must be known. This value can be obtained accurately by means of a curve tracer.

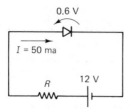

Figure 2–8 Silicon diode circuit fur current limitation.

2–4 TRANSISTORS AND SIMPLE CIRCUITS

Many different types of transistors are available. In this section the bipolar junction transistor (BJT) is discussed and the junction field effect transistor (JFET) and metal-oxide-semiconductor (MOS) are briefly described. The BJT is referred to as such because two types of charge carriers are responsible for its properties. It has two back-to-back *pn* junctions as schematically depicted in Fig. 2–9 for an *npn* transistor.

The regions of the BJT are designated E for emitter, B for base, and C for collector. The base region is very thin, approximately 1 μm. The depletion regions are shown shaded. The doping levels become progressively lighter going from emitter to collector with typical values of 10^{19} cm^{-3}, 10^{17} cm^{-3}, and 10^{15} cm^{-3} for emitter, base, and collector, respectively. The transistor symbol and an instructive, although impractical, circuit are shown in Fig. 2–10.

Referring to the circuit in Fig. 2–10, the emitter-base junction is forward biased, thus causing a large flow of electrons into the base region with a relatively small flow of holes into the emitter because of the lighter doping in the base. The base region is so thin that most electrons pass directly through it into the electric field of the reverse-biased base-collector junction, which carries them on into the collector. A small reverse saturation current occurs at the base-collector junction. The few electrons that do not make it through the base,

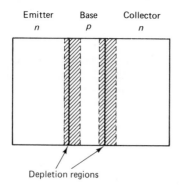

Figure 2–9 Schematic of *npn* transistor.

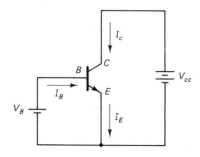

Figure 2–10 Simple circuit illustrating BJT biasing.

plus the hole current from the base to emitter, are made up for by a small flow of electrons out of the base. The currents shown in Fig. 2–10 are conventional currents; so they are in the opposite direction of the electron flow. According to Kirchhoff's current law, the three currents must be related as follows:

$$I_E = I_B + I_C \tag{2-4}$$

The dc or small-signal current gain of the transistor is $\beta = I_C/I_B$ so that

$$I_E = \left(1 + \frac{1}{\beta}\right)I_C \tag{2-5}$$

Because β usually has a value of 50 to 400, it is frequently sufficient to assume $I_E = I_C$. pnp transistors behave similarly except that the current directions and voltage polarities are reversed. Whenever possible, transistors are made npn to take advantage of better conductivity of electrons compared to holes.

Most silicon BJTs are constructed in a thin epitaxial layer on a silicon chip and so the actual device looks more like the sketch in Fig. 2–11, although it is still highly schematic. Contact to E, B, and C is made by means of aluminum deposited through openings in a SiO_2 passivation layer, which is grown directly on the Si. The n^+ symbol represents doping approaching the solid solubility limit. The buried layer is used to reduce the resistance (collector resistance) between the collector and collector contact. The drawing, of course, is not to scale. The lateral dimensions are reduced by at least a factor of ten compared to the vertical direction. Obviously some transistors, particularly high-power devices, may look much different than Fig. 2–11, but this diagram is fairly representative of BJTs used in integrated circuits, low-power applications, and optoelectronic devices.

Transistor circuits are covered thoroughly in many excellent books. It may be helpful to the uninitiated reader, however, to present some simple circuits here.

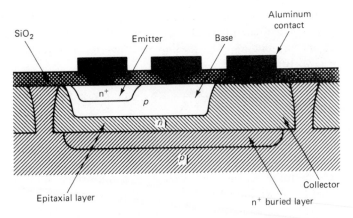

Figure 2–11   Silicon BJT.

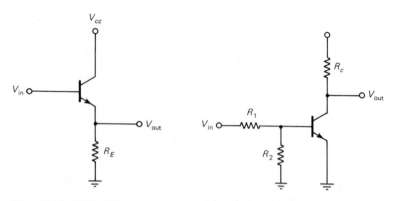

Figure 2–12 BJT used as a switch.   **Figure 2–13** BJT used as an inverter.

The BJT may be used either as an amplifier or as a switch. A simple switching scheme is shown in Fig. 2–12. The transistor does not turn on until the voltage drop across the *E-B* junction approaches 0.6 volt. In Fig. 2–12 $V_{in} = 0.6$ volt $+ I_E R$ and $V_{out} = I_E R$. Another useful configuration in logic circuits is the inverter shown in Fig. 2–13. When $V_{in} = 5$ volt, V_{out} is near zero volts and when $V_{in} = 0$, $V_{out} = 5$ volt. R_1 can be used to limit the base current and R_2 drains stored charge out of the base when the transistor is turned off to improve the speed of operation. The bleed-off resistor technique has been supplanted in integrated circuits through the use of Schottky diode clamped transistors. The Schottky diode is a metal-silicon diode that turns on at about 0.3 volt and is built from collector to base in such a way that the *B-C* junction cannot become forward biased, which is what causes charge to be stored in the base when the transistor is turned on.

Figure 2–14 contains two simple biasing schemes for small-signal, one-stage amplification. These circuits are frequently useful with optoelectronic devices. In Fig. 2–14(a), the voltage gain is $v_{out}/v_{in} = -R_c/r_e$, where $r_e = 0.026$ volt/I_E and in Fig. 2–14(b) the voltage gain is $v_{out}/v_{in} = -R_c/R_E$. The minus sign is used because of the phase change seen in the inverter and v_{out} and v_{in} are small-signal ac voltages. The biasing resistors R_1 and R_2 are chosen to make $I_1 \approx I_2 \geq 10 I_B$ so that

$$V_B = V_{cc} \frac{R_2}{R_1 + R_2} = 0.6 \text{ volt} + I_E R_E$$

where I_1, I_2, and I_E are dc currents. If the output is taken from the emitter, the amplifier is referred to as an emitter follower. The voltage gain is slightly less than one, but the output resistance is extremely low, making the emitter follower very desirable as the output stage of a multistage amplifier, such as an operational amplifier (op amp).

The purpose of the biasing resistors R_B in Fig. 2–14(a) and R_1, R_2 in Fig. 2–14(b) is to provide a dc quiescent or operating point for the transistor.

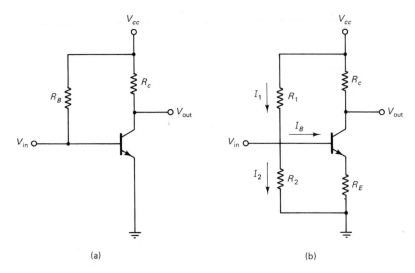

Figure 2-14   Simple common emitter amplifiers.

Then the small ac signals input to the base will be amplified at the output. One of the biasing resistors may be replaced by a photodiode or phototransistor in optical detection schemes.

A JFET is schematically depicted in Fig. 2-15. When a voltage is applied between drain D and source S as shown, conduction takes place by means of a drift current caused by the electric field produced in the channel between D and S. If V_D becomes large enough, or a negative voltage is applied to the gate G, the channel begins to pinch off as a result of an increase in the size of the depletion layer between the p and n regions. The current saturates at this point. BJTs are essentially linear in that when properly biased, changes in output current are linearly proportional to changes in input current. JFETs obey a quadratic relationship—in this case, $I_D \propto V_G^2$, where V_G is voltage applied to the gate. This nonlinearity makes FETs less desirable for linear amplifiers, but it has many useful applications in communications and optical detection.

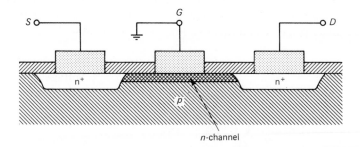

Figure 2-15 n-channel JFET.

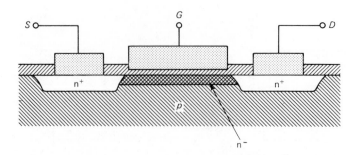

Figure 2–16   MOS n-channel depletion mode device.

JFETs have high input resistances, for the gate current is always looking into a reverse-biased diode, relative to the drain. That is, $I_s = I_G + I_D$ and I_G is extremely small. This makes JFETs useful in the first stage of high-input resistance amplifiers, such as some of the better op amps and the first amplifier stage in radios.

A MOSFET device is shown in Fig. 2–16. In these transistors conduction is by field effect—that is, drift current—just as in JFETs. The gate, however, is completely insulated from the semiconductor. These devices have also been dubbed insulated gate FETs or IGFETs. In the version shown in Fig. 2–16 a negative voltage applied to the gate pushes electrons out of the channel, thus depleting it, and this normally on device is turned off at high enough negative gate voltage. Like JFETs, the current voltage relation is quadratic. The input resistance of these devices is extremely high, because the gate is insulated. MOS devices are used heavily in digital circuits and in some optical devices.

Because of its usefulness in many optical applications, a brief introduction to the ideal op amp is given here. The reader is referred to any of a large number of electronics books for more detailed coverage. Some simple optical applications of op amps will be presented in a later chapter.

The ideal op amp has infinite input resistance, zero output resistance, and infinite open-loop gain. To a first approximation, any amplifier can be modeled by the circuit shown in Fig. 2–17. The ideal amplifier would then have $R_{in} = \infty$, $R_0 = 0$, and $A_0 = \infty$. This means that v_{in} could come from a high resistance source without losing voltage to the source resistance and the output of the amplifier can be connected to a low resistance load without losing voltage in R_0.

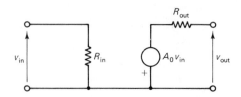

Figure 2–17   Circuit model for amplifier.

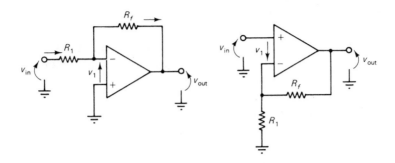

Figure 2-18   (a) Inverting amplifier. (b) Noninverting amplifier.

Having $v_0 = Av_i = \infty$ is a convenience that will become apparent as we look at some simple op amp circuits. The symbol used to represent an op amp and two useful circuits, an inverting amplifier and a noninverting amplifier, are presented in Fig. 2-18. Because of the large open-loop gain (this means no feedback loop) of the op amp, v_1 in Fig. 2-18(a) is nearly zero, thus making the negative or inverting input a "virtual" ground. Also, because of the high input resistance, the current between the $+$ and $-$ terminals is essentially zero. Consequently,

$$i_1 = \frac{v_{in} - 0}{R_1} = \frac{0 - v_{out}}{R_f}$$

or

$$A_c = \frac{v_{out}}{v_{in}} = -\frac{R_f}{R_1} \tag{2-6}$$

which is the so-called closed-loop gain. The input resistance of this amplifier is R_1 and its output resistance is[3]

$$\frac{R_0}{A_0} \frac{R_1 + R_f}{R_1}$$

In Fig. 2-18(b) the roles of the inverting and noninverting inputs are reversed. Now the voltage at the inverting terminal is approximately v_{in}. Thus

$$i = \frac{v_{in} - 0}{R_{in}} = \frac{v_{out}}{R_{in} + R_f} \tag{2-7}$$

and so

$$A_c = \frac{v_{out}}{v_{in}} = \frac{R_1 + R_f}{R_1}$$

[3] Do not confuse op amp input and output resistance with amplifier input and output resistance.

The gain is a little higher and the input and output voltages are in phase. The input resistance of this amplifier is $A_0 R_{in} (R_1/R_1 + R_f)$, which is generally very large, and its output resistance is the same as the inverting amplifier.

The input voltage in these circuits may come from a photodetector or all or part of the input resistance may be replaced by a photoresistor.

A standard low-cost op amp, the 741, has $R_{in} = 1$ to 2 MΩ, $R_0 = 75$ Ω, and a dc open-loop voltage gain of -2×10^5 that decreases with frequency to -100 at 10 kHz. In order for the relations just given to be valid, the open-loop gain of the op amp must always be much greater than the closed-loop gain.

PROBLEMS

2–1. The energy of the ground-state electron in hydrogen is -13.6 eV. In the first excited state the energy is -3.4 eV. Calculate the frequency and wavelength of the light given off when an electron undergoes a transition from the first excited state to the ground state.

2–2. Write out the electronic configurations in spectroscopic notation for Mg, Si, P, and K.

2–3. A crystal of n-type silicon at room temperature has 5×10^{18} phosphorus atoms per cubic centimeter. How many electrons and holes are there per cubic centimeter?

2–4. The turn-on voltage for a GaAs light-emitting diode (LED) is 1.2 volts. What resistance is needed in series with the LED to provide a current of 50 mA if a 12-volt battery is used?

2–5. A transistor has a β of 200. Calculate the emitter current if the collector current is 1.5 mA. What is the base current?

2–6. Design a simple inverting op amp circuit to have a closed-loop gain of 15.

2–7. Design a noninverting op amp circuit to have a gain of 10.

2–8. The energy gap in GaAs is 1.4 eV. Calculate the wavelength and frequency of the photon given off when an electron from the conduction band recombines with a hole from the valence band.

BIBLIOGRAPHY

2–1. VanVlack, L. H., *Elements of Materials Science and Engineering* (4th ed.). Reading, MA: Addison-Wesley Publishing Co., 1980.

2–2. Kittel, C., *Solid State Physics* (5th ed.). New York: John Wiley & Sons, Inc., 1976.

2–3. Till, W. C., and J. T. Luxon, *Integrated Circuits: Materials, Devices, Fabrication.* Englewood Cliffs, NJ: Prentice-Hall, Inc., 1982.

2–4. Hewlett-Packard, *Optoelectronic Designer's Catalog.* Palo Alto, CA: 1978.

chapter 3

Radiometry, Photometry, Optical Device Parameters, and Specifications

This chapter introduces radiometric and photometric quantities and the parameters and physical characteristics of solid-state light emitters and detectors that are used to describe the physical properties most frequently considered when deciding which device is best suited for a given application. The spectral output, power and/or brightness, spatial distribution of light, current requirements and limitations, and voltage requirements are discussed for light emitters. Responsivity R, detectivity D, detectivity per unit frequency D^*, noise equivalent power NEP, spectral response, and frequency response are covered for photodetectors. Some comments about noise considerations in photodetectors are made, but an in-depth coverage of this subject is beyond the scope of this book.

3-1 RADIOMETRY AND PHOTOMETRY

When the response of the eye is considered in the measurement or specification of visible radiation, the subject is called *photometry* as opposed to *radiometry*, in which direct physical measurements or specifications are made. The eye response curve is given in Fig. 3-1 for what is called photopic or day vision. The curve is shifted toward shorter wavelengths for night or scotopic vision. Not everyone, of course, has the same eye spectral response, but Fig. 3-1 is accepted as representative. The unit of luminous flux is the lumen, corresponding to the radiometric unit, the watt. Incident power per unit area (watts/cm²) in radiometric terms is called irradiance. In photometric terms it is referred to as illuminance (lumens/cm²). The units given here are not necessarily SI units

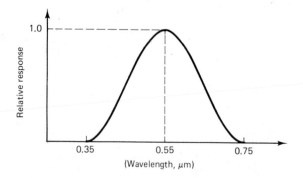

Figure 3–1   Spectral response curve of the eye (color response.)

but are commonly used combinations. The proper SI units adhere strictly to the MKS system where applicable, such as in area units.

Before discussing the remaining photometric and radiometric quantities, the concept of solid angle must be introduced. The solid angle Ω is the three-dimensional counterpart of the usual planar or two-dimensional angle defined by the ratio of the arc of a circle to the radius of the circle as illustrated in Fig. 3–2. A solid angle is defined as the ratio of the area of a portion of a sphere to the radius squared. This concept is illustrated in Fig. 3–3. The full planar angle is $2\pi r/r = 2\pi$ radians whereas the full solid angle is $4\pi r^2/r^2 = 4\pi$ steradians. A useful relation between θ and Ω exists for small angles and circular beams. Refer to Fig. 3–4 for an explanation of this relationship. As can be seen, $\Omega \approx (4\pi)\theta^2$ in this case.

Luminous intensity and radiant intensity are defined as luminous flux per unit solid angle and power per unit solid angle, respectively. A lumen per steradian is called a candela. These quantities define the ratio of the flux emanating from a source into a cone, defined by solid angle Ω to the solid angle of

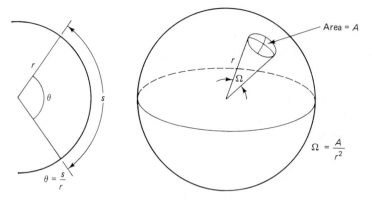

Figure 3–2 Defini-
tion of planar angle.

Figure 3–3 Definition of solid angle.

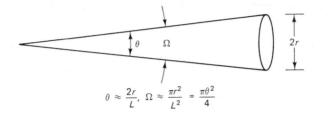

$$\theta \approx \frac{2r}{L}, \; \Omega \approx \frac{\pi r^2}{L^2} = \frac{\pi \theta^2}{4}$$

Figure 3–4   Relation between θ^2 and Ω for small-divergence angle, circular cross-sectional beam.

that cone. A laser emits a beam of very high intensity compared with most thermal light sources because the beam it emits has an extremely small divergence angle. The radiant intensity of a 1-mW HeNe laser is 1000 W/sr for a full-angle beam divergence of 1 mrad.

The final quantity of interest is intensity per unit-emitting area. In photometry it is called luminance and might have units of candelas/cm². The radiometric term is radiance and the units are frequently watts/cm²-sr. Luminance (radiance) is the flux emitted by an area of the source divided by the emitting area and the solid angle of the cone into which the flux is emitted. Luminance is often referred to as brightness.

Consider a 1-mW HeNe laser with a full-angle divergence of 1 mrad and a beam diameter of 1 mm. The radiance for this laser will be approximately 1.6×10^5 W/cm²-sr. It is informative to compare it with the radiance of the sun, which is about 130 W/cm²-sr, and a 1000-W mercury arc lamp, which has a radiance of 1000 W/cm²-sr [Ready 1971]. The radiance (luminance) of a source cannot be increased by any optical means; it can only be decreased by stops or attenuation.

3–2 PROPERTIES OF LIGHT EMITTERS

Certain *pn*-junction diodes emit light when forward biased. They are referred to as light-emitting diodes (LEDs). Like any other diode, a certain forward "turn-on" voltage must be provided and the current through the device can be limited by a series resistor. LED manufacturers specify the "turn-on" voltage and maximum allowable current. Light output versus current is nearly linear for LEDs and this curve may be supplied with the other specifications. LEDs cannot tolerate a very large reverse voltage because of low voltage breakdown due to heavy doping. The tolerable reverse voltage is given in the specifications.

LEDs emit relatively monochromatic, although incoherent, radiation with wavelength bandwidths of 100 Å. Typical wavelengths are 0.9 μm, 0.66 μm, and 0.55 μm in the ir, red, and green, respectively. The power output in watts for infrared LEDs is usually given. The spatial distribution of the power output

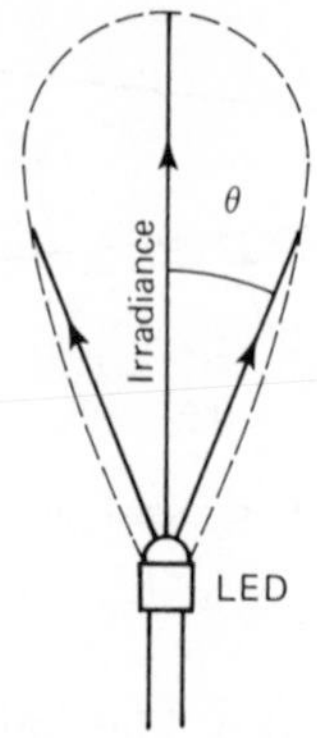

Figure 3–5 Irradiance versus angle for an LED.

may be important and is given in terms of a plot of irradiance versus angle, such as in Fig. 3–5.

The output of visible LEDs may be given in terms of intensity or brightness. Many LEDs, particularly those used in displays, approximate lambertian sources. A lambertian source or diffuser emits radiation with an intensity variation given by

$$I = I_0 \cos \theta \tag{3-1}$$

where I_0 is the intensity normal to the surface and θ is the angle between the radiating direction and the normal. Figure 3–6 depicts this situation. A reflecting or transmitting surface of this sort is considered an ideal diffuser of light. Brightness (radiance) is constant for any viewing angle because the projected area of the surface as viewed from any direction decreases as $\cos \theta$. This factor is important for displays intended for wide-angle viewing. It is left as an exercise to show that $I_0 = \pi^{-1} \times$ (total flux from the surface) for a lambertian surface. So for an approximate lambertian emitter, any two of the three—I_0, flux F, or brightness B—can be calculated if the third is given.

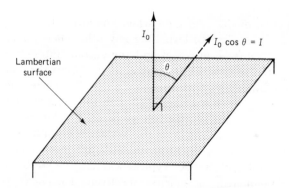

Figure 3–6 Variation of I with angle for a Lambertian surface.

3-3 QUALITY FACTORS FOR PHOTODETECTORS

Solid-state photodetectors that convert incident radiation into a measurable electrical current or voltage are compared by the use of several defined parameters. The simplest is the responsivity R, which is the ratio of the output signal (in amperes or volts) to the incident irradiance. The responsivity of a photodiode, for example, might be 1 μa/mW/cm^2 compared to 1 ma/mW/cm^2 for a phototransistor. The spectral responsivity R_λ is the responsivity per unit wavelength and is generally plotted as a function of wavelength to provide the spectral response of the device or the peak value is given along with the wavelength at which it is maximum. For photodiodes, the quantum efficiency η is sometimes given. It is simply the number of EHPs generated per incident photon. Responsivity is proportional to quantum efficiency.

The detectivity of a device, D, is the ratio of the responsivity to the noise current (voltage) produced in the detector. Thus it is the signal-to-noise (SN) ratio divided by irradiance. The units, therefore, are simply cm^2/W. This parameter is little used but serves to introduce the following parameter, D^*, which is the detectivity times square root noise frequency bandwidth, $(\Delta f)^{1/2}$, times square root detector area, $A^{1/2}$. The reason for defining D^* in this way

$$D^* = D(A\,\Delta f)^{1/2} \tag{3-2}$$

is to eliminate the dependence on noise bandwidth and detector area. Noise power is proportional to the frequency bandwidth and detector area and so the noise current or voltage will be proportional to $(A\,\Delta f)^{1/2}$. The units of D^* are cm^3 Hz$^{1/2}$/W or cm Hz$^{1/2}$/W if power instead of irradiance is used in the definition of R. D^*_λ, the spectral form of D^*, may be given in addition to D^* or in place of it.

Noise equivalent power (NEP) is the reciprocal of D and is therefore simply the irradiance required to produce a signal-to-noise ratio of one. The most widely accepted figure of merit is D^*.

The values of any parameters mentioned depend on how they are measured. Measurement conditions are usually specified. For example, $D^*(500°K, 400, 1)$ means that D^* was measured with a 500°K blackbody source chopped at 400 Hz, using 1 Hz bandwidth. Actually, the bandwidth is larger but is normalized to 1 Hz.

Noise considerations.   The main sources of noise in semiconductor photodetectors are thermal noise and generation-recombination noise. Random fluctuations in thermal motion of electrons will cause an rms current, given by

$$i_n = \sqrt{\frac{4kT\,\Delta f}{R}} \tag{3-3}$$

The random fluctuations in generation and recombinations of EHPs cause an rms current, given by

$$i_n = \sqrt{4eIG\,\Delta f} \tag{3-4}$$

where G is the photoconductive gain, which will be discussed in a later chapter.

A variety of noise sources is loosely referred to as one over f noise because all are proportional to $\sqrt{\Delta f}/f$ and hence are only important at low frequency. Because noise powers add directly, the total noise current is given by the square root of the sums of the squares of the individual noise currents. Note that in all cases the noise current is proportional to $(\Delta f)^{1/2}$.

Sources of radiation are also sources of noise. At low light levels statistical variations in the rate of arrival of photons at the detector cause what is called quantum noise. Fluctuations in the interference of waves are a source of thermal noise in a detector. Another source of noise arises from randomly varying capacitance due to vibrations or breakdown of dielectric films.

3-4 SPECTRAL RESPONSE

The spectral response is generally presented as a variation of responsivity as a function of wavelength. Spectral response characteristics for specific types of detectors are described in Chapter 4, but some general comments are appropriate here. Photodetectors fall into two general categories—thermal detectors and quantum detectors. Examples of thermal detectors are calorimeters, thermopiles, bolometers, thermistors, and pyroelectric detectors. In all these devices the incident radiation is converted to heat, which, in turn, causes a temperature rise and a corresponding measurable change, such as an emf generated in a thermocouple or a current pulse in a pyroelectric detector. Detectors of this type tend to have a flat spectral response over a wide wavelength range.

Examples of quantum detectors are photomultiplier tubes, semiconductor photodiodes, and phototransistors. In this type of detector photons produce a measurable change directly, such as the emission of electrons in photomultiplier tubes or the generation of EHPs in semiconductor photodiodes. Photoemissive surfaces for photomultipliers have been designed with fairly flat spectral response but have a sharp cutoff at the long wavelength end where the photon energy is insufficient to overcome the work function of the material. Semiconductor detectors have fairly narrow spectral response, peaked at the wavelength corresponding to photons with energy equal to the band gap energy.

3-5 FREQUENCY RESPONSE

The frequency response of a photodetector refers to its ability to respond to an incident light beam chopped or modulated at various frequencies. Most solid-state detectors behave like lowpass filters and as such their responsivity can be written

$$R = \frac{R_0}{(1 + \omega^2\tau^2)^{1/2}}$$

NA·/Rad.

where R_0 is the responsivity at zero frequency and τ is the ti
The cutoff frequency is $f_c = (2\pi\tau)^{-1}$, where $R = R_0/\sqrt{2}$. The t
may be given by $\tau = rC$, where C is the device capacitance and
resistance plus circuit resistance. Clearly an inherently fast devic
quency response) can be slowed down by use of a large circuit resi ce.

PROBLEMS

3–1. Calculate the solid angle for a HeNe laser that emits a beam with a full-angle divergence of 1 mrad if the diameter of the emitted beam is 2 mm.

3–2. The total flux (power) from a lambertian surface is 2 mW. Calculate the intensity at normal incidence and the brightness (radiance) if the surface is 1 mm square.

3–3. Calculate the noise current in a 100-Ω resistor at room temperature for a bandwidth of 1 kHz.

3–4. The dc responsivity of a solid-state photodetector is 10 mA/mW with a time constant of 10^{-7} s. What is the responsivity at 10^8 Hz?

3–5. Name and briefly discuss the reasons for three types of noise common to photodetectors.

3–6. Name the figures of merit used to qualify photodetectors. Write out the equations for them and explain.

REFERENCE

READY, J. F., *Effects of High-Power Radiation*. New York: Academic Press, 1971.

BIBLIOGRAPHY

3–1. Stimson, A., *Photometry and Radiometry for Engineers*. New York: John Wiley & Sons, Inc., 1974.

3–2. Williams, W. S., and O. A. Becklund, *Optics: A Short Course for Engineers and Scientists*. New York: John Wiley & Sons, Inc., 1972.

3–3. Driscoll, W. G. (ed.), and W. Vaughn (assoc. ed.), *Handbook of Optics*. New York: McGraw-Hill Book Co., 1978.

3–4. Chappel, A., (ed.), *Optoelectronics: Theory and Practice*. The Netherlands: Holland Printing Partners. Texas Instruments Ltd., 1976.

3–5. RCA, *Electro-Optics Handbook*. Harrison, NJ: RCA Commercial Engineering, 1968.

chapter 4

Light Detection
Devices

This chapter describes a variety of light detection devices that are frequently used in laser applications. The discussion includes energy and power measurement as well as a brief description of imaging devices, such as photodiode arrays and charge-coupled devices (CCDs). Some simple detector circuits are also briefly described.

4–1 CLASSIFICATION OF DETECTORS

Light detectors can be placed in either of two classifications: thermal or quantum (sometimes referred to as photon). Thermal detectors are those in which the absorbed light energy is converted into heat that produces a temperature rise that, in turn, causes some detectable output or variation in a property of the detector. Quantum detectors are those in which the absorbed light directly alters a property or produces a measurable output.

Examples of thermal detectors are thermocouples, calorimeters, bolometers, and pyroelectric devices. The radiation causes a resistance change in bolometers; either a voltage or a current pulse is produced in the others. In all cases, it is a change in temperature that produces the measurable phenomenon.

Quantum detectors are exemplified by photoemissive devices, such as photomultiplier tubes (PMTs), photodiodes, photoconductors, and phototransistors. A resistance change is directly produced in photoconductors by the absorption of light; in the others a measurable current or voltage is generated by direct interaction of the absorbed light with electrons.

Generally thermal detectors are slower and have a broader spectral response than quantum detectors. There are notable exceptions to this statement, however, particularly with respect to the speed of response.

4–2 THERMAL DETECTORS

As noted, thermal detectors are based on a change in property or output that results from a temperature increase arising from the absorption of light. In most devices a certain degree of "thermal inertia" is inherent because heat conduction is involved. Consequently, the response time of such detectors tends to be quite slow unless the mass is kept small, as in some thin film devices. Most thermal detectors use a blackened absorbing surface and frequently exhibit relatively flat spectral response from the uv to the far ir. Window materials placed in front of the absorbing surface and the absorbing coating determine spectral response.

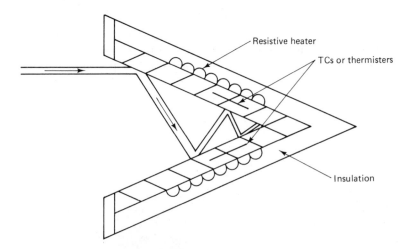

Resistive heater

TCs or thermisters

Insulation

Figure 4–1   Cone-type calorimeter.

Calorimeters.   Although many different types of calorimeters exist, only a few are briefly described here to elucidate the principles involved.

Several calorimeter designs use an absorbing cavity that is suitably coated for high absorption and may have a conical shape to produce multiple reflections of the light entering the cone. The temperature change is sensed by thermocouples or thermistors[1] embedded in the absorber, which is usually aluminum or copper because of their large thermal conductivity. Such a device is schematically represented in Fig. 4–1.

---

[1] A thermistor is a ceramic resistor with a large temperature coefficient of resistance.

When light is absorbed by such a device for a period of time, the absorbed energy causes a uniform temperature increase in the highly conductive Al or Cu. This temperature increase is sensed by the thermocouples or thermistors and may be displayed as a voltage on a meter or digital readout. Calibration is carried out by irradiation with a standard source or by supplying a known quantity of heat through a resistive heater embedded in or attached to the absorber. Such calorimeters can measure power by being allowed to reach thermal equilibrium with their surroundings or coolant. The detected steady-state temperature rise is then a measure of the incident power. Calibration would occur in the same way as for energy measurement.

Because the response time of these devices may be several seconds, it is possible to measure high power levels by chopping the beam and allowing only a small fixed percentage (e.g., 1 to 3%) of the power to enter the calorimeter. In addition, because of the inherent "thermal inertia" of such devices, the output signal will remain relatively steady. Power levels of thousands of watts can be measured in this fashion. Figure 4–2 is a photograph of a commercial calorimeter that uses an absorption cavity for high-power CO_2 laser power and/or energy measurement.

Some calorimeters are water cooled and the difference in the inlet and outlet water temperature is used as a measure of the power or energy absorbed. Because heat capacity C is defined as heat absorbed per unit mass per unit

Figure 4–2 Absorption cavity calorimeter. (Courtesy of Apollo Lasers, an Allied Company.)

temperature change, Eq. (4–1) gives the power P absorbed in terms of heat capacity, temperature change ΔT, and mass flow rate dm/dt:

$$P = \frac{dQ}{dt} = C\,\Delta T \frac{dm}{dt} \qquad (4\text{–}1)$$

A "rat's nest" calorimeter uses wire, such as copper, randomly tangled inside a highly reflective container. The tangled mass (or mess if you prefer) of wires ensures nearly 100% absorption of the incident radiation and the temperature rise is sensed as an increase in resistance of the wire. Absolute calibration is achieved by dissipating a known amount of electrical power in the wire. Because a thermally induced resistance change is used, this device could also be called a rat's nest bolometer.

When thermistors or bolometers are used to sense temperature changes, it is possible to place two identical calorimeters in a bridge arrangement for very sensitive power or energy measurements. Ambient conditions are the same for both calorimeters except that one is irradiated and the other is not. This technique is depicted in Fig. 4–3. The variable resistance is adjusted for a null reading on the galvanometer. When the calorimeter is illuminated, the bridge is unbalanced, thus causing a deflection of the galvanometer that can be calibrated in terms of power or energy for maximum deflection. The identical dummy calorimeter tends to provide cancellation of resistance change effects due to variations in environmental conditions, which should nominally be the same for both calorimeters.

Thermopiles.   A thermopile consists of a number of thermocouples connected in series. A thermocouple is a junction of two different metals or semiconductors. Because of the work function difference between the materials, a current is generated when the junction is heated. Thermocouples use a reference junction and the voltage difference between the junctions is a measure of the temperature difference between them. The reference voltage in modern systems

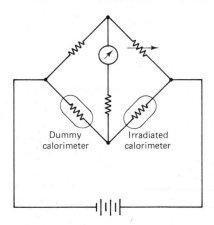

Figure 4–3 Bridge network for calorimeter.

is established electronically so that only one junction, or probe, is required. Such devices are extremely common in temperature measurement. When a junction of dissimilar materials, with an appropriate absorbing coating, is irradiated, a temperature increase occurs, thus generating a voltage that is a measure of the power absorbed.

The voltage produced in metallic thermocouples is only a few microvolts per degree change in temperature and so it is common practice to arrange several couples in series so that all the junctions of one polarity are irradiated. In this way, their voltages add to give a larger response, the reference junctions being masked from the radiation. Figure 4–4 is a sketch of such a thermopile. The thermopile may be incorporated as one leg of a bridge circuit with an identical dummy device in the other leg to improve sensitivity and immunity to environmental fluctuations. Thin film thermopiles can have response times of the order of microseconds with a flat spectral response.

Bolometers.   Bolometers were discussed earlier and so little will be added here. Such devices are either metallic or semiconductor resistors in which a thermal change in resistance is induced by absorbed radiation. The change in resistance causes a change in current in a circuit and hence a voltage change across the bolometer or another resistor in the circuit. Bolometers, of course, can be used in a bridge circuit with a dummy bolometer in one leg.

Pyroelectric detectors.   Pyroelectric detectors are made from materials that have permanent electric dipole moments associated with them, such as $LiNbO_3$, triglycine sulfate (TGS), and polyvinylfluoride (PVF). These materials are ferroelectric, meaning that permanent dipole moments are associated with the unit cell, in the case of crystalline materials, or the basic molecular unit in the case of polymers. These dipoles are all aligned parallel within domains analogous to the domains in ferromagnetic materials. The chief difference between pyroelectric ferroelectrics and other ferroelectrics is the difficulty of poling pyroelectrics. In order to get the saturated polarization of all domains aligned

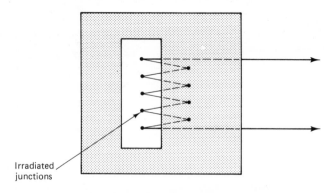

Irradiated
junctions

Figure 4–4   Thermopile.

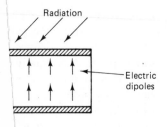

Radiation

Electric dipoles

‑oelectric detector scheme.

t be placed in a strong electric field at ization cannot be reversed at normal ric field. The total dipole moment of sensitive to temperature fluctuations. detector, light is converted to heat, ence a change in polarization. Figure ʜeme. Charges absorbed on the surface When the polarization is altered due ansferred through the external circuit, e quickly. The speed of response for ond range, because the speed is not ..., (ʜeʀmaɪ inertia) limited. The temperature change causes a change in dipole moment that responds extremely fast.

In the circuit of Fig. 4–5 R should be less than the resistance of the detector, which is quite high because it is basically a capacitor. The voltage developed by the detector is given by

$$V(t) = AR\gamma \frac{dT}{dt} \qquad (4-2)$$

where γ is the pyroelectric coefficient or polarization change per unit temperature change and A is the irradiated area.

Pyroelectric detectors have a flat spectral response to long wavelengths without the need for cryogenic cooling required by other ir detectors and their frequency response, which obeys Eq. (3–5), is very good with bandwidths around a gigahertz. Maximum incident power is limited to a few tens of watts per square centimeter.

4–3 QUANTUM DETECTORS

Devices that produce a response as a direct result of the absorption of radiation are referred to as *quantum* or *photon detectors* and, indeed, some photoemissive devices can detect single photons.

Photoconductors.   When a photon is absorbed by a semiconductor, the probability that an electron-hole pair will be produced is quite high. The quantum efficiency η, which is defined as the ratio of conduction electrons produced to absorbed photons, is usually 0.7 or higher at the peak of the spectral response curve. A photoconductor is an intrinsic or extrinsic semiconductor in which the change in conductivity resulting from the absorption of light is used as a measure of the incident power. Unlike a bolometer, temperature change has nothing to do with the conductivity change. Conductivity is given by

$$\sigma = ne\mu_e + pe\mu_h \qquad (4\text{--}3)$$

where n and p are the number of electrons and holes per unit volume, respectively, and μ_e and μ_h are the mobilities of electrons and holes, respectively. The creation of electron-hole pairs in an intrinsic conductor alters both n and p. The long wavelength cutoff of the device is given by

$$\lambda_c = \frac{hc}{E_g} \qquad (4\text{--}4)$$

where E_g is the width of the energy gap, h is Planck's constant, and c is the speed of light. The signal current, assuming the dark resistance is very large, is given by

$$i_s = e\eta A \frac{\Delta f t_c}{t_t} \qquad (4\text{--}5)$$

where A is the irradiated area, Δf is the bandwidth, t_c is the carrier lifetime, and t_t is the transit time (time for an electron to traverse the device). The ratio t_c/t_t is called the photoconductive gain because i_s can be increased by designing the photoconductor to have long carrier lifetime and short transit time. The main source of noise in photoconductors is generation-recombination noise, which also increases with t_c/t_t; so the signal-to-noise ratio is not improved by increasing the photoconductive gain.

Narrow bandgap photoconductors, such as HgCdTe, are used extensively for ir detection but must be cryogenically cooled to minimize the effect of thermally generated charge carriers.

Photodiodes.   Semiconductor diodes were discussed in Chapter 3 and so the basic theory of diodes need not be considered here. When a semiconductor diode is used as a photodetector, an internal current is produced as a result of the absorption of photons and the creation of electron-hole pairs (EHPs). Many types of photodiodes and modes of operation are possible. The more common types are discussed next.

All pn-junction devices respond to incident radiation if the junction is close enough to the surface for light to reach it and if the wavelength of the incident radiation is less than the cutoff wavelength. Longer wavelengths tend

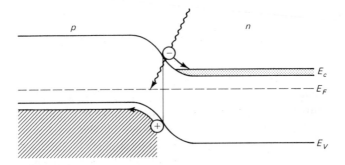

Figure 4-6   Energy level diagram for an unbiased photodiode.

to penetrate farther and short wavelengths are absorbed nearer the surface. The uv response of photodiodes is poor both because EHPs generated near the surface tend to encounter surface recombination centers and because most of the energy of the uv photons is wasted, for only a small fraction is needed to produce an EHP.

The process by which current is generated can be understood by reference to Fig. 4-6. It can clearly be seen that an EHP generated in the depletion region will tend to produce a current as the election and hole slide down their respective potential hills. If the diode has a resistor connected across it, a current is produced as long as light is incident on the junction. This situation is depicted in Fig. 4-7. This is the photovoltaic mode of operation, for a voltage is produced across the resistor and power $V_d I_d$ is produced.

Actually, EHPs need not be generated exactly in the depletion region. EHPs generated within one diffusion length on either side of the junction will contribute to the current. The diffusion length is the average distance that an electron or hole travels in the minority regions before recombining. The diffusion length for holes is substantially less than that for electrons; so most diffused diodes are made with a thin p-type region on top (toward the incident radiation).

The ideal responsivity of photodiodes is easily deduced in the following way. The photocurrent is given by

$$I_p = \eta e \frac{dn}{dt}$$

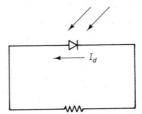

Figure 4-7   Photovoltaic mode of operation.

where dn/dt, the rate at which photons strike the detector, is given by P_i/hf. Here P_i is the incident power. The responsivity in amperes per watt is then

$$R = \frac{I_p}{P_i} = \frac{e\lambda}{hc} = 8.1 \times 10^5\lambda \ (A/W) \tag{4-6}$$

which yields a typical responsivity of about 0.6 A/W at $\lambda = 1.0$ μm.

Photodiodes may have areas of 1 cm² or greater for high sensitivity or position detection, but this factor increases noise and capacitance. Many photodiodes are very small, having an active area of less than 0.01 cm². The responsivity for small-area devices is more conveniently defined as output current per unit irradiance in watts/cm². A typical responsivity is 1 mA/W/cm².

The major source of noise in photodiodes is shot noise, which is caused by the tendency of electrons and holes to "bunch up" or to cross the depletion region in a random manner.

The short wavelength response of photodiodes can be enhanced by placing the junction closer to the surface. Overall response is then sacrificed, of course. To achieve improved response at all wavelengths in the spectral range of the device, *pin* diodes are used. Here an intrinsic layer (actually a very lightly doped region) is placed between the *p* and *n* regions. This step improves quantum efficiency by providing a larger depletion region and increases speed of response by decreasing junction capacitance.

Two common silicon photodiode designs are the planar-diffused and Schottky barrier diodes. Figure 4–8 contains sketches of both types. The planar-diffused type may be made as a *pin* or ordinary *pn* diode, depending on the spectral response and speed requirements. The Schottky barrier diode has a superior uv response, due to the fact that the depletion region extends to the silicon surface. Large-area Schottky diodes are less noisy than their planar-diffused counterparts but have lower power-handling capability.

The diode equation can be generalized to account for photogenerated current. If a short circuit is placed across an illuminated diode, the resulting diode current is a reverse current and is called the short-circuit current, I_{sc}. If a resistor is placed across the illuminated diode, then a forward-bias voltage is produced and the diode current is given by

$$I_d = I_0(e^{eV/kT} - 1) - I_{sc} \tag{4-7}$$

Note that the open-circuit voltage, V_{oc}, is given by

$$V_{oc} = \frac{kT}{e}\left(\ln\frac{I_{sc}}{I_0} + 1\right) \tag{4-8}$$

Figure 4–9 is a sketch of diode current versus diode voltage for various incident light levels. When a diode is operated in the photovoltaic mode, it is operating in the fourth quadrant of the current versus voltage graph. The short-circuit current varies nearly linearly with incident power over about six decades. So

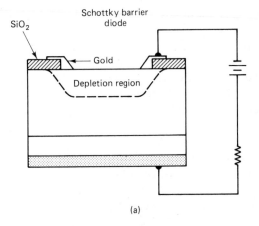

(a)

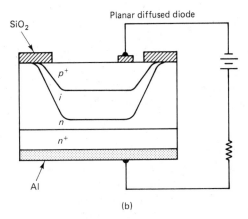

(b)

Figure 4-8 Silicon photodiode designs.

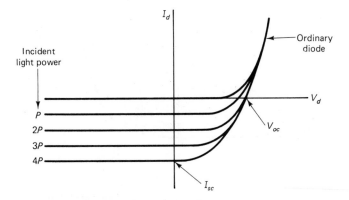

Figure 4-9 Diode current versus diode voltage for various light levels.

a circuit that provides essentially zero voltage drop across the diode provides substantially linear operation.

Maximum power output in the photovoltaic mode, $(V_d I_d)_{max}$, is given by a point determined by inscribing a rectangle of maximum area as shown in Fig. 4–10. The process involves matching the load resistance to the diode resistance. Unfortunately, the diode resistance varies with incident light level in the photovoltaic mode. Consequently, no single resistance will work for all light levels and an optimum load based on operating conditions must be determined. Solar cells and nonbattery-powered light meters operate in this fashion.

When a photodiode is reverse biased, the operation may be linear over 10 or 11 orders of magnitude change in light level and the junction capacitance is reduced, thereby decreasing the response time in some diodes to less than 1.0 ns. Reverse-bias operation is referred to as the photoconductive mode, but it should not be confused with the term photoconductor; photoconductors are not even diodes.

Avalanche photodiodes are reverse biased at a precisely controlled voltage that is slightly less than the breakdown voltage. When light strikes the diode, creating EHPs, the avalanche process begins and continues as long as the light is incident on the diode. The advantage of this type of diode is the multiplication factor. It may be as great as 1000, which provides an internal gain and correspondingly high responsivity. The response time of avalance diodes is also very good, about 1.0 ns.

Phototransistors.   A phototransistor is just what the name implies— a transistor used as a light detector to take advantage of the internal gain of the transistor. Light incident on the base-collector junction of a bipolar transistor produces an internal current that acts as the base current of an ordinary transistor. Consequently, no base lead is required unless external biasing is necessary.

The spectral response of silicon phototransistors is similar to that of silicon

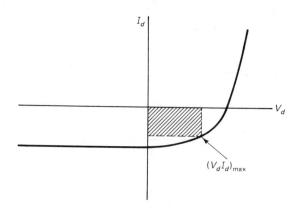

Figure 4–10   Maximum power output point.

diodes, but their responsivities are about three orders of magnitude larger, typically around 1 A/W/cm². The response time of phototransistors is microseconds and slower.

Many phototransistors are built in a Darlington configuration to take advantage of the increased current gain realized from such a design. The light is allowed to impinge on the collector-base junction of the first transistor and the second one provides additional gain. The responsivity of these devices is typically 10 A/W/cm² or higher.

Photo-FET devices are available for applications in which the very high input resistance of such transistors is required. Photo-SCRs are also available for light-activated switching applications, such as fire alarms.

Photon drag.   Detectors using the principle of photon momentum transfer, referred to as *photon drag detectors*, are not extensively used in manufacturing settings, but they are potentially useful for on-line monitoring of CO_2 laser power and measurement of the time dependence of CO_2 laser pulses.

The momentum of a photon can be determined by equating the mass-energy equivalence relation to the photon energy

$$mc^2 = hf$$

so that momentum, $p = mc$, is given by

$$p = \frac{hf}{c} = \frac{h}{\lambda} \qquad (4\text{–}9)$$

In such materials as germanium and tellurium, which are transparent to 10.6-μm radiation, a sufficient amount of momentum is transferred from the photons to electrons or holes to cause a measurable open-circuit voltage across a crystal through which the radiation is passing. This open-circuit voltage is the result of the holes or electrons being pushed or "dragged" along with the radiation. Response time may be less than 1.0 ns with responsivities of 10^{-8} to 10^{-6} V/W.

Photomultipliers.   If the energy of photons incident on an absorbing surface exceeds the work function[2] of the electrons in that surface, the photoelectric effect is observed. The incident photons eject electrons from the surface by direct transfer of energy. Collection of these electrons at an anode produces a photoelectric current that is a measure of the light power incident on the photoemissive cathode.

Photodetectors that use a single photoemissive surface and a single collector are simply called *phototubes* or *photodiode tubes*. Phototubes are quite fast but have low responsivities because there is no internal gain mechanism. In

---

[2] Work function is the energy required to remove an electron from the surface of the material to infinity.

photomultipliers a series of "dynodes" are used to provide multiplication or gain. The electrons ejected from the cathode are accelerated by a voltage of 100 to 200 V to a dynode where they, on impact, eject many secondary electrons, which are, in turn, accelerated to another dynode and so on. Tubes may contain ten or more dynodes and produce a multiplication factor of one million or more. Response times (transit time of the electrons) of less than 1.0 ns are achieved and single photons can be detected, thus leading to the phrase photon counting even though not every photon incident on the emissive surface is detected.

A variety of photocathodes covers the spectral range from 0.1 to 11.0 μm. Examples are (a) Ga-As, which has a fairly flat spectral response from 0.13 to 0.8 μm with a peak responsivity of around 35 mA/W, and (b) Ag-O-Cs alloy, which covers the spectral range from 0.3 to 1.0 μm, although the responsivity varies significantly.

Many photomultiplier designs exist. Figure 4–11 is a schematic representation of one type of design. Photomultiplier tubes are extremely fast and sensitive but expensive and they require sophisticated associated electronics to operate.

Position detectors. A variety of silicon photodiode position detection devices is available. Such devices fall into two categories: multidiode units and lateral-effect diodes.

Multidiode units consist of either two or four separate, but identical, photodiodes placed on the same silicon wafer but electronically isolated from each other. The bicell and quadrant detectors are schematically depicted in Fig. 4–12. When a symmetrical laser beam is centered on either detector, the output of each diode is identical. If the beam moves more onto one diode than the other(s), the signal from that diode increases while the signal from the other(s) decreases. This change in output can be used to measure the displacement of the laser beam. Measurements of beam displacement of as little as 1.0 μm are claimed to be possible. One-directional displacements can be measured with the bicell whereas two-dimensional displacements can be measured with quadrant

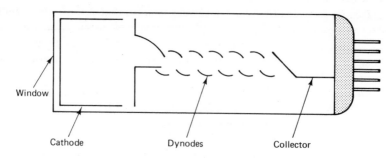

Figure 4–11   Photomultiplier tube.

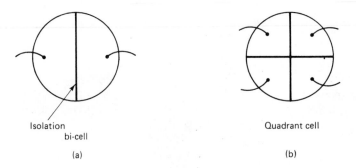

Isolation

bi-cell

(a)

Quadrant cell

(b)

Figure 4-12   Photodiode position detectors.

detectors. The linear displacement of the beam on the detector can also be easily related to angular displacement of the beam.

The beam profile generally used with position detectors is a filtered Gaussian distribution. Consequently, the output of the diodes is not linearly related to the amount of beam displacement. Detectors of this type can be made either by the planar-diffused process as either *pn* or *pin* diodes or by the Schottky barrier process.

The lateral-effect diode position detector avoids the problem of nonlinear output as a function of beam displacement. Figure 4-13 is a schematic diagram of a lateral-effect diode. The output current from terminal S is ideally given by

$$I_s = I_0\left(1 - \frac{S}{L}\right) \tag{4-10}$$

where I_0 is the actual photogenerated current across the depletion layer at the point of incidence of the beam. The farther the beam is from the output

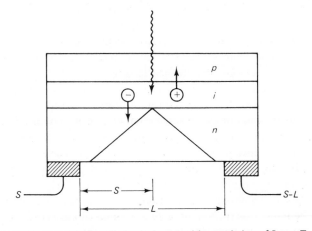

Figure 4-13   Lateral-effect diode. (Adapted with permission of Laser Focus.)

Figure 4–14   Silicon photodetectors.

terminal, the more electrons cross back over the depletion layer and recombine at the surface. This type of position detector is slower and noisier than multicell types, but the change in output with beam position is very linear and the beam irradiance profile is irrelevant. Beam displacements as small as 0.01 μm can be detected with silicon diode position detectors and this factor allows very small angular changes to be detected.

Figure 4–14 shows a variety of silicon photodetectors that can be used for power measurement, centering, and position measurements.

4–4 DETECTOR ARRAYS

A thorough discussion of detector arrays is well beyond the scope of this book. A brief description of the basic types of arrays is included, however, because of their usefulness in lower power laser applications. Linear and area arrays using photodiodes, charge injection devices (CIDs), and charge-coupled devices (CCDs) are available. Detector arrays are frequently used in systems designed to measure part dimensions and surface location.

Photodiode arrays.   As the name implies, photodiode arrays consist of either one-dimensional or two-dimensional arrays of photodiodes. The output of each diode is proportional to the amount of light incident on it; so imaging is possible with a resolution dependent on the cell (pixel) size and center spacing of the pixels. In this context, a pixel refers to the basic light detection unit.

The light-sensitive area is about half the total pixel area. These devices are usually back illuminated so that there are no contact obstructions. Figure 4–15 is a schematic illustration of a linear photodiode array. Arrays of this type may have from 64 to 2048 elements on center-to-center spacings as low

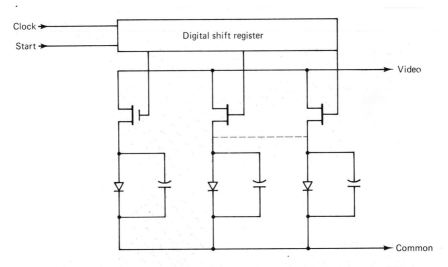

Figure 4–15   Linear photodiode array. (Adapted courtesy of EG & G Reticon Corp.)

as 2.5 μm. Light incident on the silicon photodiodes produces charge which is stored in the parallel capacitor until a voltage from the shift register is applied to the gate of the FET. The charge is then dumped onto the video line so that a sequence of pulses whose amplitudes are proportional to the amount of light energy incident on the diode during the framing period is read out. The slower the clock rate, the more charge can be stored and the more sensitive the device is to light level. Sampling rates may be as high as 40 MHz. These devices also come with CCD-analog shift registers. In this version a voltage is applied simultaneously to the gates of all the FETs, the charge stored in each capacitor is dumped into the CCD shift register, and the charge packets are then read out serially as with the digital shift register. In both types of arrays the output is a discrete time analog of the light irradiance distribution across the array.

Two-dimensional diode arrays with up to 256 × 256 elements are available. In these devices a shift register sequentially applies pulses to the FET gates row by row dumping the charge from each diode in a row into one or two CCD shift registers, which then serially output the charge packets onto one or two video lines. Pixel rates of 10 MHz can be achieved.

Charge-coupled devices.   CCDs use metal-oxide-semiconductor (MOS) capacitors directly to store packets of charge produced by the incident light. The charge packets, which represent an integration of the light incident on a particular capacitor, are then serially shifted out of the CCD by transference from one capacitor to another. Simplified versions of two schemes for achieving CCD imaging are shown in Fig. 4–16.

In three-phase CCD imagers the charge is collected under the gates of

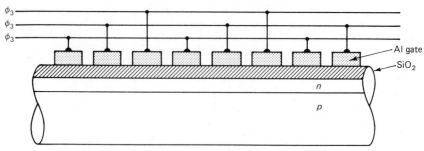

(a) Three-phase charge-coupled device

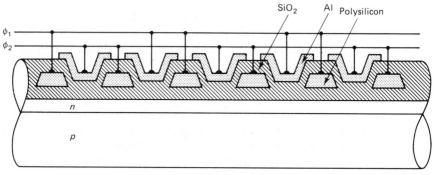

(b) Two-phase charge-coupled device

Figure 4–16   CCD linear imagers. (Adapted with permission form *Solid State Technology*, Technical Publishing, a company of Dun & Bradstreet.)

capacitors that have a negative gate voltage applied—for instance, θ_1. As a result, a depletion region is created under the θ_1 gate oxide into which holes generated by incident photons are attracted and stored. The number of holes in the packet is proportional to the number of photons incident during the period of time that θ_1 is negative. The charge packets can then be transferred to the θ_2 capacitors by reducing θ_1 and making θ_2 large and then to capacitor θ_3 by making θ_1 and θ_2 small and θ_3 large and so on until the charge is read out as an analog pulse similarly to the output of a photodiode array. Alternatively, the packets stored in the imaging capacitors can be transferred initially to a parallel line of MOS capacitors that are masked so that light cannot reach them and then the charge is transferred to the output in the manner described.

To prevent the trapping of electrons by fast surface states at the oxide interface, a "fat zero" (a small number of holes) can be cycled through the capacitors continually to keep these states filled. This process minimizes the degradation of the signal. Another technique is to use ion implantation to put a shallow p-type buried layer below the oxide; doing so pushes the depletion layer down into the bulk Si away from the surface states and keeps the signal charge away from the surface state traps.

Charge transfer in the right direction in three-phase devices is determined by the asymmetry in the application of the clock voltages, θ_1, θ_2, and θ_3. Charge transfer in the right direction can also be accomplished in a two-phase system by introducing a built-in asymmetry. Figure 4–14(b) illustrates one way of doing so. The oxide thickness under the poly-Si gates is less than under the Al gates; thus the oxide capacitance for the poly-Si gate capacitors is higher than that of the Al gate capacitors. Consequently, a smaller voltage occurs in the depletion region beneath the Al gate than below the poly-Si gate capacitors. This situation can be considered in terms of potential wells: the wells under the Al gates are always less deep than the wells under the corresponding poly-Si gates. If θ_1 is large, charge is accumulated in the θ_1 capacitors. As θ_1 is decreased and θ_2 increased, charge can only move to the right because of the potential barrier that exists on the left due to the shallower well under the Al gate.

Transfer efficiencies, the amount of charge transferred from one MOS capacitor to the next, of over 99.999% are possible. Center-to-center spacing of pixels is typically 12.7 μm and a dynamic range[3] of 1000 to 1 is common. The elements in CCD imagers can be made about as small as the limits of photo or electron beam lithography will allow. Very large linear and area arrays that can operate at both high- and low-light level conditions are possible. It has been reported that 200 million pixels per second can be read in large displays through the use of multiple readout taps.

Charge injection devices.   CID imagers are similar to CCDs in that photon-generated charge is collected and stored in MOS capacitors. In two-dimensional imagers each pixel consists of two side-by-side capacitors. When a gate voltage is applied to one, it gathers and stores charge. The voltage is then transferred to the second capacitor, which is masked from the light, by application of the voltage to it and removal of the voltage from the first. Readout is accomplished by sequentially setting the voltages to zero for a line that has been selected by placing all the second (storage) capacitor voltages high. When the voltage goes to zero, the charge is injected into the substrate, thus causing a detectable current. Using this technique, it is possible to address a two-dimensional array on an individual basis (random access) by applying appropriate column and row voltages. A linear array does not require the second set of capacitors unless individual addressing is necessary.

4-5 PHOTODETECTOR CIRCUITS

Circuits for the operation and amplification of photomultiplier tubes, photodiode arrays, and CCD or CID arrays are beyond the scope of this book. Such circuitry

---

[3] Ratio of highest to lowest detectable light levels.

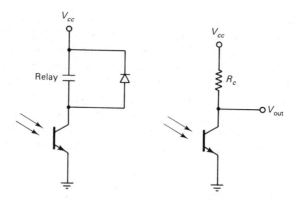

Figure 4–17 Light-activated relay.

Figure 4–18 Phototransistor amplifier.

can be found in the applications notes provided by the manufacturer. A few simple but useful circuits for photodiodes, phototransistors, or photoconductors are described, however.

A simple light-activated relay circuit is depicted in Fig. 4–17. Phototransistors can handle sufficient current to activate small relays. When light strikes the phototransistor, turning it on, the relay is opened or closed, depending on its normal state. When the light is removed, the relay returns to its normal state. The diode may be placed in parallel with the relay to protect it from possible voltage spikes.

A simple phototransistor circuit of the sort shown in Fig. 4–18 can be used for switching or detection operations and even for light-modulation applications in which true fidelity of the modulation wave form is not critical. Voice communication on an amplitude-modulated laser beam, for example, can be

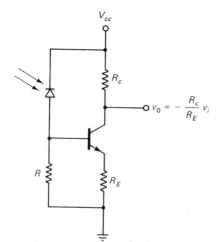

Figure 4–19 Transistor amplifier circuit for a photodiode.

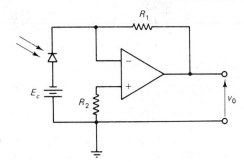

Figure 4-20 Op-amp circuit for linear photodiode operation.

accomplished by taking the output of this phototransistor amplifier and feeding it into an audio amplifier.

A transistor circuit that can be used to amplify a photodiode output is given in Fig. 4–19. The photodiode in Fig. 4–19 may be replaced by a photoconductor.

An operational amplifier circuit for photodiodes that provides great linearity and high speed is depicted in Fig. 4–20. In this circuit E_c and R_2 may be zero. E_c is provided to reverse bias the diode, which produces faster and more linear operation. The resistance R_1 may be chosen to balance offset current resulting from diode dark current. The output voltage for this circuit is given by

$$v_0 = -i_p R_1 \qquad (4-11)$$

In applications where large light level changes may be encountered, a logarithmic amplifier, such as the one given in Fig. 4–21, may be useful. In this case, the op amp must have an extremely high input impedance. It can be obtained by using an FET input stage. Here the op amp is used in a noninverting amplifier arrangement where

$$v_0 = \left(1 + \frac{R_1}{R_2}\right) v_i$$

Because of the very high input impedance, $I_d \simeq 0$ so that

$$0 = I_0(e^{eV_d/kT} - 1) - I_{sc}$$

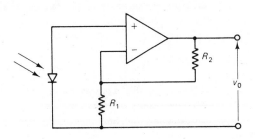

Figure 4-21 Logarithmic amplifier for photodiode applications.

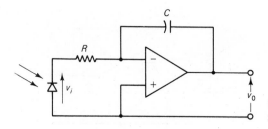

Figure 4–22　Integrating circuit for energy measurement.

which leads to

$$V_d = v_i = \frac{kT}{e} \ln\left(1 + \frac{I_{sc}}{I_0}\right)$$

Thus the output voltage is given by

$$v_0 = \left(1 + \frac{R_1}{R_2}\right) \frac{kT}{e} \ln\left(1 + \frac{I_{sc}}{I_0}\right) \tag{4–12}$$

If energy measurement is desirable, a simple op amp integrating circuit (shown in Fig. 4–22) may serve the purpose. In this circuit the output voltage is proportional to the integral of the input voltage, which is, in turn, proportional to the power incident on the photodiode. Therefore the output voltage is a measure of the energy absorbed by the diode during the time interval of the integration. Equation (4–13) is the relation between output voltage and input voltage.

$$v_0 = \frac{1}{RC} \int_0^t v_i \, dt \tag{4–13}$$

PROBLEMS

4–1. (a) Name the two classes of photodetectors, describe them, and explain how they work.

 (b) List three types of detector in each class and explain why they are classified that way.

4–2. What should the mass flow rate in a water-cooled calorimeter be if 1C° temperature change is to occur for each 100 W of power absorbed?

4–3. Assuming the specific heat of a pyroelectric detector is C and its mass m, determine an expression relating output voltage, area, series resistance, pyroelectric coefficient, the given quantities, and the power P incident on the device.

Answer: $\quad V = AR\gamma \dfrac{P}{mc}$

This assumes all the power is absorbed by the detector and none is lost by conduction, convection, or radiation.

4–4. (a) Calculate the conductivity of an n-type silicon with $N = 2 \times 10^{18}$ cm^{-3}

chapter **5**

Introduction to the Laser

The purpose of this chapter is to present an overall view of lasers and their applications. In this way, it is hoped that the reader will have a clearer and more cohesive understanding of the material presented in subsequent chapters. It may also enable the reader to be selective, if desired, in the choice of reading from the remaining chapters. The reader who has some familiarity with lasers and the various types available may wish to skip this chapter.

The acronym L A S E R is derived from the expression Light Amplification by Stimulated Emission of Radiation. Used in this sense, the term refers to electromagnetic radiation anywhere in the spectrum from the ultraviolet (uv) through the infrared (ir), which includes wavelengths from roughly 0.1 to 1000 μm. The term M A S E R was originally coined to describe a similar device (using ammonia molecular transitions and a microwave cavity) that operates in the microwave region of the electromagnetic spectrum—in this case, at a wavelength of 1.25 cm.

The acronym, laser, although slightly redundant, is highly descriptive. Light and radiation, in this context, are the same thing. The key words are "amplification" and "stimulated emission." It is the ability of light to stimulate the emission of light that creates the situation in which light can be amplified. It may be helpful to point out at this time that a very important feature of most lasers is an optical resonator, usually consisting of two precisely aligned mirrors, one of which is partially transmitting, to allow an output. This mirror arrangement provides positive feedback. So the laser is basically a positive feedback oscillator.[1] As such, it is analogous to electrical positive feedback oscillators where a certain amount of the output is fed back in phase with the input, resulting in oscillation at some frequency characteristic of the circuit. In effect, the oscillator selects a frequency component from the noise always present from biasing, amplifies it, and oscillates at that frequency. The laser does essentially the same thing except that an optical oscillator can operate in many allowed modes (natural resonator frequencies).

5-1 UNIQUE PROPERTIES OF LASER LIGHT

The laser is basically a light source. The radiation that it emits is not fundamentally different than any other form of electromagnetic radiation. The nature of the device, however, is such that some remarkable properties of light are realized. These unique properties, taken as a whole, are not available from any other light source to the extent that they are obtained from a laser. The unique properties referred to are

1. High monochromaticity (small wavelength spread)
2. High degree of both spatial and temporal coherence (strong correlation in phase)
3. High brightness (primarily due to small beam divergence)
4. Capability of very low (microwatts) to very high (kilowatts) continuous power output for different types of lasers
5. High peak power (terrawatts) and large energy (hundreds of joules) per pulse in pulsed output lasers
6. Capability of being focused to a small diffraction limited spot size (of the order of the wavelength of the light)

These properties are by no means independent of each other and, in fact, some may be inferred directly from others.

5-2 REQUIREMENTS FOR LASER ACTION

A number of conditions must be satisfied to achieve lasing action. They are listed here and are briefly discussed at this point with more detail following in later chapters.

---

[1] In some cases, such as the N_2 laser, the gain is sufficiently great that feedback is not required to achieve lasing action.

1. Population inversion
2. Optical resonator, except in extremely high gain systems
3. Lasing medium
4. Means of excitation
5. Host medium (usually)

The notion of a population inversion refers to a condition in which a certain ensemble of atoms or molecules is in a nonequilibrium situation where more of these atoms or molecules are in some specified excited energy state (electronic or vibrational) than are in a lower energy state. These atoms or molecules undergo a transition to a lower state in which the probability of emission of a photon is extremely high, a so-called radiative transition.

The optical resonator refers to the technique for providing positive feedback into the system to produce oscillation. This process usually consists of two parallel mirrors placed some distance apart; one is as nearly totally reflecting as possible and the other is partially transmitting to obtain a useful output from the system. The transmittance of the output mirror, which is the ratio of transmitted power to incident power, ranges roughly from 1 to 60%, depending on the power level and type of laser.

The lasing medium (called the lasant) refers to the atoms or molecules that actually emit the light, such as Ne atoms in a HeNe Laser or Cr^{3+} ions in a ruby laser.

Some means of excitation is required to achieve the population inversion. This step is usually accomplished by the electrical discharge or high intensity light from gas-discharge lamps, such as xenon or krypton gas-discharge lamps. Many other important techniques for excitation exist, however, such as chemical reaction, electron beam preionization, nuclear and gas dynamic. In the chemical reaction pumped lasers, two or more reactive gases are mixed in a chamber where an exothermic reaction occurs, thus, producing the energy required for the population inversion and the lasant as well. In electron beam preionization a broad beam of electrons is accelerated through a high voltage into a foil from which many more slower electrons are ejected; such electrons then pass through the laser cavity, causing ionization by collision. Excitation is actually achieved by means of an electrical discharge at a voltage too low to sustain ionization but that is more effective at achieving excitation. Nuclear excitation refers to the use of a beam of particles from a nuclear reactor or accelerator to cause excitation. Gas dynamic lasers use differences in decay rates from excited states to achieve a population inversion when a high temperature gas is forced through supersonic nozzles.

A host medium is one in which the lasant is dispersed. In the ruby laser Al_2O_3 is the host and serves as a matrix to hold the Cr^{3+} ions. Helium is the host in the HeNe laser and is essential to the process of exciting the lasant,

Ne. Some laser types have no host; examples are the semiconductor diode laser and ion lasers like argon and krypton.

5-3 HOW THE LASER WORKS

This section provides a brief qualitative explanation of how lasers work. More detailed information is presented in later chapters.

All laser action begins with the establishment of a population inversion by the excitation process. Photons are spontaneously emitted in all directions. Photons traveling through the active medium can *stimulate* excited atoms or molecules to undergo radiative transitions when the photons pass near the atoms or molecules. This factor in itself is unimportant except that the stimulated and stimulating photons are in phase, travel in the same direction, and have the same polarization. This phenomenon provides for the possibility of gain or *amplification*. Only those photons traveling nearly parallel to the axis of the resonator will pass through a substantial portion of the active medium. A percentage of these photons will be fed back (reflected) into the active region, thus ensuring a large buildup of stimulated radiation, much more than the spontaneous radiation at the same frequency. Lasing will continue as long as the population inversion is maintained above some threshold level.

The optical resonator, because of its geometrical configuration, provides for a highly unidirectional output and at the same time, through the feedback process, for sufficient stimulated radiation in the laser to ensure that most transitions are stimulated. The phenomenon of stimulated emission, in turn, produces a highly monochromatic, highly coherent output (some exceptions will be discussed later). The combined action of the resonator and stimulated emission produces an extremely bright light source even for lasers of relatively low power output.

5-4 TYPES OF LASERS

No attempt will be made to list all the significant laser types commercially available. Most industrial lasers will be briefly described here with additional detail presented on selected lasers in a later chapter. The lasers to be discussed at this point are HeNe, ruby, Nd-YAG, Nd-Glass, CO_2, dye, and the semiconductor diode.

The HeNe laser is by far the most common laser in use today. It is a gas laser with He the host and Ne the lasant. The output wavelength is 0.6328 μm (red) for most applications, although it can emit at two ir wavelengths and power levels range from microwatts to about 50 mW continuous. This is a highly reliable, low cost laser and is used extensively in such applications as alignment, surveying, ranging, displacement measurement, holography, pattern recognition, communications, surface finish analysis, and flow measurement. The means of excitation in the HeNe laser is a dc glow discharge. The gas atoms that participate in the lasing process are not ionized; so the HeNe laser is a neutral gas laser.

The ruby laser has Al_2O_3 (sapphire) as the host and triply ionized chromium ions, Cr^{3+}, as the lasant. The output wavelength is 0.6943 μm (red). It is operated in pulsed fashion with a low repetition rate because of poor thermal properties of Al_2O_3. Many joules of energy can be realized in a single pulse, however. This energy has been used for drilling diamonds, sending pulses to the moon, spot welding, hole piercing, and pulsed holography. Excitation is by means of capacitive discharge through flash lamps.

The Nd-YAG and Nd-Glass lasers are similar in that both have triply ionized neodymium (knee-oh-dim-e-um) atoms (Nd^{3+}) as the lasant. YAG stands for yttrium-aluminum-garnet, a synthetic crystal also used for simulated diamonds. Its chemical formula is $Y_3Al_5O_{12}$. The output wavelength for both laser types is 1.06 μm (near ir). Nd-YAG lasers are operated continuously or at high repetition rates with average or continuous power output of from a few watts to 1 kW. Applications of Nd-YAG lasers include welding, cutting, hole piercing, and excitation of dye lasers with frequency quadrupling. Nd-Glass lasers are pulsed because of the poor thermal characteristics of glass, but extremely high peak powers have been achieved. Lasers of this type are pumped by gas-discharge lamps. Some applications for Nd-Glass lasers are welding, cutting, hole piercing, and laser fusion experiments.

There are several different basic CO_2 laser designs, but, in general, they use CO_2 as the lasant and also contain N_2 and He. The emission wavelength is 10.6 μm, which is in the mid infrared, and the power output is continuous or pulsed with average power ranging from a few watts to tens of kilowatts. Excitation techniques include dc glow discharge, rf, electron beam preionization, chemical and gas dynamic. As far as industrial lasers are concerned, glow discharge, rf, and electron beam preionization excitation are of greatest interest. Applications for this laser are numerous and include cutting, hole piercing, welding, heat treatment, scribing, and marking. The variety of materials that can be worked is also varied, including paper, plastic, wood, glass, cloth, ceramics, and most metals.

Dye lasers are devices in which the lasant is an organic dye. Because organic dyes fluoresce in a large number of lines (distinct wavelengths), the laser can be tuned to emit at a variety of wavelengths over some wavelength band. By using several dyes, it is possible to obtain thousands of distinct, closely spaced[2] wavelengths in the visible part of the spectrum from a single laser with fairly uniform power output at all these wavelengths. Power outputs range

---

[2] Line broadening causes the discrete lines to overlap, thus, in effect, producing a continuously tunable laser.

from milliwatts to watts. These lasers are pumped by xenon lamps, argon lasers, nitrogen lasers, excimer lasers, or frequency-multiplied Nd-YAG lasers. The output of dye lasers is normally pulsed. Dye lasers are used in spectroscopy, photochemical reaction studies, pollution detection, and surgery.

The semiconductor diode laser is a distant cousin of the light-emitting diode (LED) in that it is a semiconductor diode that emits light. The diode laser, however, emits coherent radiation with a much narrower wavelength spread than the incoherent emission of an LED. Modern diode lasers are chiefly composed of a variety of ternary compounds,[3] such as $GaAl_xAs_{1-x}$. Feedback in diode lasers is achieved either by cleaving the ends of the diode parallel to each other and perpendicular to the diode junction to form mirror surfaces or by distributed feedback via a diffraction grating etched on top of the chip parallel to the junction. The light is emitted approximately parallel to the junction rather than roughly at right angles to it as in most LEDs. Physically diode lasers are small, being on the order of 250 μm on a side by 50 μm thick. These devices, however, can emit peak powers of many watts in 100- to 200-ns pulses with an average power output of several milliwatts. Diodes operating continuously at room temperature can emit several milliwatts of power. Wavelengths vary slightly for $GaAl_xAs_{1-x}$ but are close to 0.9 μm (near ir). Applications for diode lasers appear to be in fiber optic communications, pattern recognition, ir illumination, and pollution detection and control.

PROBLEMS

5-1. List six unique properties of laser light.

5-2. List five requirements for lasing action to occur and briefly explain what each means.

5-3. List four important industrial lasers and suggest a possible type of application for each, along with the output wavelength and the range of CW power or energy per pulse.

5-4. Starting with the emission of a spontaneous photon along the axis of a laser, qualitatively describe the lasing action in a CW laser. Use sketches to illustrate your discussion and describe the relationship of the radiation to the initial photon.

5-5. What is the function of the mirrors in a laser?

BIBLIOGRAPHY

5-1. O'Shea, D. C., W. R. Callen, and W. T. Rhodes, *Introduction to Lasers and Applications*. Reading, MA: Addison-Wesley Publishing Co., 1978.

5-2. Lengyel, B. A., *Introduction to Laser Physics*. New York: John Wiley & Sons, Inc., 1966.

---

[3] A compound containing three elements in contrast to a binary compound, which contains two.

chapter 6

Laser Theory

The objectives of this chapter are to provide a sufficient theoretical basis so that readers can pursue additional study on their own and appreciate properly the material presented in later chapters. A basic understanding of what the laser is, how it works, and what its unique properties are will help dispel the mystery that surrounds this useful device. When understood, tools are much more frequently used and used correctly.

Only a brief theoretical treatment can be presented in this limited space. Many results are given without proof and derivations are kept as simple as possible without loss of meaning.

6-1 POPULATION INVERSION

One requirement for lasing action is a population inversion. Establishing a population inversion requires that more lasant atoms or molecules be in some excited state than in the state to which the most probable transition occurs. This transition must be one in which photons are emitted (radiative transition) as opposed to one in which no photons are emitted (nonradiative transition). In the latter case, the energy ends up as thermal instead of electromagnetic energy.

If only two energy levels were involved, the system would be called a two-level system. The only laser that actually resembles this situation is the diode laser[1] since the transitions take place between levels near the band edges

---

[1] The ammonia maser is a two-level system.

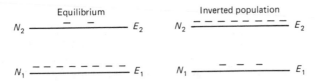

Figure 6-1   Two-level energy level scheme.

of the semiconductor. Most lasers are better approximated by three- or four-level systems. The actual energy level structure of atoms or molecules is much more complicated than that, however. Nevertheless, much can be learned by considering a hypothetical two-level system. Some discussion of three- and four-level lasers will be presented subsequently. With this in mind, we initially discuss population inversion, plus a number of other matters, in the context of a two-level system.

A two-energy-level system is depicted in Fig. 6-1. Here E_1 and E_2 are the energies of the respective levels and N_1 and N_2 are the number of atoms or molecules per unit volume in each level. The ratio of N_2 to N_1 at thermal equilibrium is given by the Boltzmann equation, which is present as Eq. (6-1).

$$\frac{N_2}{N_1} = e^{-(E_2 - E_1)/kT} \tag{6-1}$$

Here k is the Boltzmann constant and T is absolute temperature. It can be seen from this relation that N_2 can never exceed N_1 for an equilibrium situation. A population inversion refers to a nonequilibrium situation in which $N_2 > N_1$. It occurs as a result of the excitation process and has been referred to as a "negative temperature" situation because Eq. (6-1) can only be made to predict $N_2 > N_1$ if T is negative. This was an attempt to fit a thermal equilibrium result to a nonequilibrium condition and is not considered further here. A population inversion is absolutely necessary for amplification of light to occur. Without a population inversion, a beam of light directed through the medium with photon energy $hf = E_2 - E_1$ will be absorbed.

6-2 STIMULATED EMISSION

Three types of transitions involving photons in laser operation are of interest: absorption, spontaneous emission, and stimulated emission. Absorption occurs when a photon with energy equal to the energy difference between two levels is absorbed, thus causing an excitation. In electronic transitions the electron is raised to a higher energy level; in molecular transitions the vibrational energy of the molecule is increased. Atoms or molecules[2] in excited states eventually

---

[2] Frequently the term atom will be used where either atom or molecule could apply.

drop to a lower energy state spontaneously; in radiative transitions a photon is given off with energy equal to the difference in the energy levels.

The precise theory of the phenomenon of stimulated emission is well beyond the scope of this book, which does not, however, preclude understanding the nature and consequences of the phenomenon. Basically what happens is that a photon passing sufficiently close to an excited atom causes it to undergo a radiative transition before it would otherwise do so spontaneously. The stimulated photons have the same phase and the same polarization state and travel in the same direction as the stimulating photons. Consequently, gain is achieved when a beam of light of the right frequency (photon energy) is directed through a medium in which a population inversion has been established.

The fact that the stimulators and stimulatees are in phase with each other produces a high degree of both spatial and temporal coherence in lasers. Usually substantially less than perfect temporal coherence is achieved because of random fluctuations that occur in the laser due to vibrations and thermal variations. Stabilization techniques can be used to increase temporal coherence greatly and hence coherence length. The lateral spatial coherence of laser beams is extremely good.

The occurrence of the phenomenon of stimulated emission is not too surprising with hindsight. The atoms or molecules behave like excited electric dipoles. When these dipoles come under the influence of the electric and magnetic field associated with a photon, the dipole experiences a force. If the phase and frequency of the dipole and photon match (also spatial orientation of the photon electric field and dipole orientation), this force results in the dipole giving off radiation (a photon). This interaction has often been referred to as "tickling" the excited atoms with the photon to produce the stimulated emission.

6–3 EINSTEIN PROBABILITY COEFFICIENTS

Based on thermodynamic arguments, Einstein established what is essentially the theoretical basis for the laser decades before the first one was built. This theory deals with the probabilities for absorption, spontaneous emission, and stimulated emission of photons. Actually, it is necessary to talk about the rates of the various types of transitions in order to get a grip on the subject. The rate of spontaneous emission can be written $N_2 A_{21}$, where N_2 is the number of atoms per unit volume in the upper state (2) and A_{21} is the probability per unit time that a given atom in an excited state will spontaneously undergo a transition to state 1. The product $N_2 A_{21}$ then gives the number of spontaneous transitions per unit volume per unit time (rate per unit volume). The absorption and stimulated rates are a little harder to describe because neither can occur if no radiation field is present. It is reasonable to expect these rates to be proportional to the radiation energy density. Thus the product $N_1 \rho_f B_{12}$ gives the rate of absorption of photons per unit volume. The energy density ρ_f is a spectral quantity and therefore has the units joule/m³Hz. B_{12} is the probability per unit time per unit spectral energy density that a given atom will be excited from state 1 to state 2. The units of B_{12} are complicated because of the absorption rate dependence on ρ_f; they are m³Hz/joule-s. The rate of stimulated emission can be expressed as $N_2\rho_f B_{21}$, where N_2 and ρ_f are the same as previously introduced. B_{21} has the same interpretation as B_{12} except that it relates to stimulated transitions from state 2 to state 1. At equilibrium, assuming that only optical transitions occur, these rates are related as follows:

$$N_1\rho_f B_{12} = N_2 A_{21} + N_2\rho_f B_{21} \tag{6-2}$$

This equation simply states that the rate of upward transitions (absorptions) equals the rate of downward transitions (emissions) at equilibrium for a two-level system. By using the Boltzmann relationship given in Eq. (6-1), the ratio N_1/N_2 can be eliminated from Eq. (6-2), yielding

$$e^{(E_2 - E_1)/kT}\rho_f B_{12} = A_{21} + \rho_f B_{21} \tag{6-3}$$

Setting $E_2 - E_1 = hf$, the energy of a photon absorbed or emitted, and solving for ρ_f yield

$$\rho_f = \frac{A_{21}}{B_{12}e^{hf/kT} - B_{21}} \tag{6-4}$$

Prior to Einstein's work, Planck worked out a model of blackbody radiation that correctly predicted the spectral emission and radiation density of a blackbody, the ideal absorber and radiator. The spectral radiation density at frequency f, according to Planck's theory, is

$$\rho_f = \frac{8\pi hf^3}{c^3} \frac{1}{(e^{hf/kT} - 1)} \tag{6-5}$$

Because Eqs. (6-4) and (6-5) are equivalent for a two-level system, we can compare them and see what it tells us about the Einstein coefficients. The only way the right-hand sides of Eqs. (6-4) and (6-5) can be equal is if $B_{12} = B_{21}$ and A_{21} is related to B_{21} by

$$A_{21} = \left(\frac{8\pi hf^3}{c^3}\right) B_{21} \tag{6-6}$$

The conclusions reached here are extremely important. The fact that $B_{12} = B_{21}$ means that, for a given radiation density, a stimulated radiation is just as probable for an atom in state 2 as absorption is for an atom in state 1. Furthermore, the coefficient of B_{21} in Eq. (6-6) is numerically 5.6×10^{-16} (SI units) at 10^{14} Hz, which means that in a radiation field of density 10^{-6} j/m³Hz the rate of spontaneous radiation is 10^{10} times smaller than the rate of stimulated radiation. This result ensures that when a population inversion is established and some feedback mechanism is included to provide reasonable radiation density at the correct frequency, more stimulated emission than spontaneous emission will occur and the device will lase. As noted, all that is needed to achieve amplification is a population inversion and a beam of light of the correct frequency, but it generally must come from a laser.

6–4 AMPLIFICATION

If there were no losses in a medium with an inverted population, the growth in radiation would be geometrical.[3] This situation is illustrated for the usual linear laser or amplifier arrangement in Fig. 6–2. As can be seen, the first photon creates one more; these two, in turn, produce four, which, in turn, produce eight and so on. A geometrical growth implies that there will be a fixed fractional increase in the number of photons per unit length of the medium.

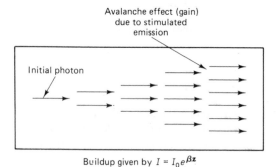

Buildup given by $I = I_0 e^{\beta z}$

Figure 6–2 Optical amplifier.

Because the irradiance is proportional to the flux of photons, this can be expressed as

$$\frac{dI}{I} = \beta \, dz$$

where I is irradiance, z distance, and β the fractional increase in irradiance per unit length or gain coefficient. Beer's law says that the loss of irradiance in a dielectric medium is given by $dI/I = -\alpha \, dz$, where α is the fraction loss per unit length. Thus the net fractional loss or gain is $dI/I = (\beta - \alpha) \, dz$, which when integrated is

$$I = I_0 e^{(\beta - \alpha)z} \tag{6–7}$$

There are various causes of losses in a laser, such as absorption and scattering in the lasing medium, absorption by windows used to seal discharge tubes, absorption by the mirrors, and, of course, the output. In one complete round

---

[3] We are no longer talking about an equilibrium situation and it should be recognized that a population inversion in a two-level system cannot be achieved by optical pumping.

trip in a laser of optical length[4] L the round trip gain, $I/I_0 = G$, is given by $R_1 R_2 e^{(\beta-\alpha)2L}$, where R_1 and R_2 are the reflectances of the mirrors and all other losses are included in α. This can be written

$$G = e^{(\beta-\alpha')2L} \qquad (6\text{–}8)$$

where

$$e^{-\alpha'2L} = R_1 R_2 e^{-\alpha 2L}$$

Henceforth it will be assumed that all losses are incorporated in α.

It is instructive to derive an expression for the gain coefficient on the basis of the population inversion and the Einstein coefficients. Assuming that spontaneous emission is negligible, we can write the net transition rate as

$$(N_2 \rho_f B_{21} - N_1 \rho_f B_{12})A \; \Delta z = (N_2 - N_1)\rho_f B_{21}A \; \Delta z$$

This transition rate can be equated to the power (irradiance times area) $\Delta I_f A$ generated in the volume element $A \; \Delta z$ divided by the energy per photon hf. Thus

$$(N_2 - N_1)\rho_f B_{21}A \; \Delta z = \frac{\Delta I_f A}{hf} \qquad (6\text{–}9)$$

In the limit as $\Delta z \to 0$ we then have

$$\frac{dI_f}{dz} = (N_2 - N_1)\rho_f B_{21}hf \qquad (6\text{–}10)$$

For an unidirectional (linear) beam, $I_f = \rho_f c$ so that Eq. (6–10) becomes

$$\frac{dI_f}{dz} = \frac{(N_2 - N_1)I_f B_{21}hf}{c} \qquad (6\text{–}11)$$

which integrates to

$$I_f = I_0 e^{[(N_2-N_1)B_{21}hf/c]z} \qquad (6\text{–}12)$$

where I_0 is some initial irradiance. Therefore

$$\beta = \frac{(N_2 - N_1)B_{21}hf}{c}$$

We now substitute for B_{21} in terms of A_{21} and get

$$\beta = \frac{(N_2 - N_1)c^2 A_{21}}{8\pi f^2} \qquad (6\text{–}13)$$

Recall that A_{21} is the probability per unit time that an excited atom will undergo a spontaneous transition. Therefore the reciprocal of A_{21} is the average time

[4] See Chapter 1 for the definition of optical length.

τ that an atom remains in the excited state before making a spontaneous transition. Now we can write the gain coefficient as

$$\beta = \frac{(N_2 - N_1)c^2}{8\pi f^2 \tau} \tag{6-14}$$

Although we have ignored line-broadening effects (discussed later) and degenerate energy levels (more than one level with the same energy), some interesting conclusions can be drawn from Eq. (6–14).

The most obvious conclusion from Eq. (6–14) is verification of the fact that $N_2 > N_1$ is required to achieve gain, if $N_2 < N_1$ loss is incurred. A second interesting point is that the gain is inversely proportional to the frequency squared (line-broadening effects make this even worse). This statement says that lower gains will be achieved at higher frequencies and therefore high-frequency lasers will probably be harder to develop than low-frequency ones. It appears to be borne out by the low number of lasers in the uv compared with visible and ir lasers and the total absence of lasers in the x-ray[5] or γ-ray region. The third very interesting point is that the gain is inversely proportional to the spontaneous lifetime τ. It is often pointed out that the upper lasing level should be metastable or a long lifetime state. It can be seen from Eq. (6–14) that a short lifetime is desirable for high gain. Because a finite amount of time is required to achieve a population inversion, a compromise must be reached. We would like τ as short as possible to provide high gain but sufficiently long to allow the population inversion to be established and maintained in the case of continuous output (CW) lasers. It is also desirable that the lifetime of the upper lasing level be substantially longer than the lifetime of the lower level, although it is not strictly necessary in pulsed lasers.

6-5 POWER OUTPUT FOR CW OPERATION

It is helpful for an understanding of both CW and pulsed laser operation to look at how CW operation relates to the gain coefficient and optical losses. If we assume some initial power P_0 directed along the axis of the laser, then the power flux at the same point after one complete round trip would be

$$P_0 + P_0 e^{(\beta - \alpha)2L}$$

where L is the optical length of the cavity. The power output of the laser would be

$$P = TP_0[1 + e^{(\beta - \alpha)2L}]$$

where $T = 1 - R$ is the transmittance of the output mirror. The TP_0 term is

---

[5] Such lasers have been reported, but the jury is still out.

retained because this power is always being generated spontaneously. After another round trip we have

$$P = TP_0[1 + e^{(\beta-\alpha)2L} + e^{2(\beta-\alpha)2L}]$$

Here it is seen that each existing term is amplified and a new TP_0 is generated. The power expression for a continuously operating laser becomes an infinite series:

$$P = TP_0 \sum_{n=0}^{\infty} e^{n(\beta-\alpha)2L} \qquad (6\text{–}15)$$

In the case where $n(\beta - \alpha)2L$ is less than zero (which it must be for stable CW operation), this series can be summed and Eq. (6–15) becomes

$$P = TP_0[1 - e^{(\beta-\alpha)2L}]^{-1} \qquad (6\text{–}16)$$

This result implies that β must be at least slightly less than α for CW laser action, which seems contrary to common sense. However, it must be recognized that if $\beta > \alpha$, then P would increase indefinitely, which is impractical for CW operation or any other kind for that matter. To have equilibrium, the optical power generated must be exactly balanced by optical losses, including the useful output. The gain coefficient β must be slightly less than α because of the small amount of spontaneously generated optical power. Stated another way, the apparent deficit due to the fact that the gain is less than the loss is compensated for by spontaneously produced power. Fortunately, the spontaneous power is negligible compared with the stimulated power in lasers.

The value of β given by $\beta_{th} = \alpha$ is called the *threshold value* of the gain coefficient. The gain must closely approach this value in order for continuous lasing action to occur.

6–6 PULSED OPERATION

In pulsed operation of lasers the gain is raised substantially above the threshold value, generally by means of flash lamps in solid lasers or pulsed electrical discharge in gas lasers. As threshold is exceeded, lasing action begins, but the population inversion continues to increase for some time, resulting in a rapidly increasing power output. This situation quickly leads to a diminution of the population inversion and a drop in power. The power output of a pulsed ruby laser looks somewhat as shown in Fig. 6–3(a). The power spiking is the result of the periodic buildup and depletion of the population inversion. The gradual decline in peak power is the result of the diminishing power output of the flash lamps. Effectively, the laser is alternately lasing and not lasing. The reason is clear from Eq. (6–14); β is alternating between values much larger than α and less than α as the population inversion alternately increases due to the optical pumping and decreases due the rapid growth in stimulated emission.

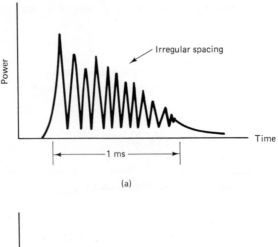

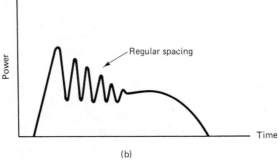

Figure 6–3   Laser pulses.

The pulsing is random in multimode[6] lasers due to mode hopping. In single-mode operation periodic oscillations occur in the first part of the pulse (relaxation oscillations) and die out to a smoothly diminishing power output for the remainder of the pulse [Fig. 6–3(b)].

6–7 THREE- AND FOUR-LEVEL SYSTEMS

In reality, lasers are more nearly approximated by three- or four-level systems, such as those schematically represented in Fig. 6–4.

Two common types of lasers are accurately represented as three-level lasers—ruby and dye lasers. In both cases, excitation is achieved by optical pumping to some level, or band of levels, E_2, followed by rapid nonradiative decay to some lower level E_3. Lasing action then takes place between level E_3 and the

---

[6] See Section 6–9 for a discussion of laser resonator modes.

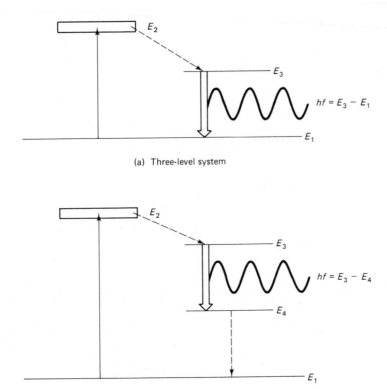

(a) Three-level system

(b) Four-level system

Figure 6–4   Three- and four-level energy level schemes.

ground state E_1. The threshold level pump energy is quite high for these lasers because the entire population of lasant species must be inverted.

In four-level lasers, such as Nd-YAG, HeNe, ion, and CO_2, excitation is to a level or band of levels E_2, followed by rapid nonradiative transition to level E_3 with lasing occurring between that level and another intermediate level E_4. It is hoped that rapid relaxation by nonradiative transition to the ground state E_1 occurs. The potential for CW operation in four-level lasers is much higher than in three-level lasers, for the threshold pump energy is drastically lower: there is no need to invert the entire population of lasant species. It is only necessary that $N_3 - N_4$ exceed some threshold value. Excitation is by optical methods in solid lasers and usually by electrical discharge in gas lasers. In CO_2, HeNe, and metal vapor lasers various types of energy transfer by collision with other ions, molecules, or atoms play a major role in producing the population inversion.

6–8 RATE EQUATIONS

Primarily for better understanding rather than practical use, the rate equations for three- and four-level lasers are presented here and discussed briefly.

The appropriate equations for the three-level system are

$$N_1 + N_3 = N_t \tag{6-17}$$

$$\frac{dN_3}{dt} = W_p N_1 - Bp(N_3 - N_1) - \frac{N_3}{\tau} \tag{6-18}$$

$$\frac{dp}{dt} = VBp(N_3 - N_1) - \frac{p}{\tau_c} \tag{6-19}$$

where $W_p N_1$ accounts for the pumping rate, τ is the lifetime of the lasing state, τ_c is the cavity lifetime of a photon, $B = B_{12} = B_{21}$ is the Einstein stimulated transition coefficient (probability rate per photon),[7] p is the number of photons in the cavity, and V is the mode volume[8] of the active region. Equation (6–17) states that the total number of atoms N_t is shared between the lasing state and the ground state; N_2 is assumed to be negligible because of the rapid decay from the pump level. Equation (6–18) expresses the rate of change of atoms in the lasing state as the pump rate minus the stimulated and spontaneous emission rates. Equation (6–19) expresses the rate of change of photons in the cavity in terms of the net stimulated emission rate minus the rate of loss of photons, p/τ_c.

Letting $\Delta N = N_3 - N_1$ represent the population inversion, Eqs. (6–17) to (6–18) can be reduced to the following two equations.

$$\frac{d\,\Delta N}{dt} = W_p(N_t - \Delta N) - 2Bp\,\Delta N - \frac{(\Delta N + N_t)}{\tau} \tag{6-20}$$

$$\frac{dp}{dt} = \left(VB\,\Delta N - \frac{1}{\tau_c} \right) p \tag{6-21}$$

Equation (6–21) tells us that lasing action will begin at some critical value of population inversion, $\Delta N_c = 1/(\tau_c VB)$. For $\Delta N > \Delta N_c$, $dp/dt > 0$ and amplification begins. A critical pump rate can be obtained from Eq. (6–20) by setting $d\,\Delta N/dt = 0$, $\Delta N = \Delta N_c$ and $p = 0$. Setting $p = 0$ is justified by the fact that the number of photons present in the cavity prior to the onset of lasing is very small. Thus

$$W_{pc} = \frac{1}{\tau}\frac{(N_t + \Delta N_c)}{(N_t - \Delta N_c)} \tag{6-22}$$

---

[7] Previously the Einstein coefficients were based on energy density.

[8] Mode volume is the volume of the active material actually filled by radiation.

Equation (6–22) reduces to

$$W_{pc}N_{1c} = \frac{N_{3c}}{\tau} \qquad (6\text{–}23)$$

using $N_t = N_{3c} + N_{1c}$ and $\Delta N_c = N_{3c} - N_{1c}$. This last equation simply states that the rate of pump transitions equals the rate of spontaneous transitions at the critical pump rate for the onset of lasing action. Actually, $N_{1c} \approx N_{3c}$ so that approximately $W_{pc} = 1/\tau$. There must, however, be a large enough population inversion to overcome the losses of the laser.

A steady-state condition for a constant pump rate that exceeds the critical pump rate will be reached where $dp/dt = 0$ and $d\,\Delta N/dt = 0$. This results in

$$\Delta N = \frac{1}{VB\tau_c} = \Delta N_c \qquad (6\text{–}24)$$

$$p = \frac{V\tau_c}{2}[W_p(N_t - \Delta N)] = \frac{(N_t + \Delta N)}{\tau} \qquad (6\text{–}25)$$

The steady-state population inversion is exactly the same as the critical inversion. Higher pump rates mean proportionately higher numbers of photons in the laser. Moreover, because the power output is directly proportional to the number of photons, the power output is also linearly proportional to the pump rate.

The output power can be calculated from $P_{\text{out}} = hfpcT/2L$. This conclusion follows from the fact that each photon has an energy hf and all the energy in the resonator strikes the output mirror in the time $2L/c$ with T giving the fraction emitted. It does not mean, of course, that more and more power output can be achieved by supplying more and more power to the laser. Clearly the pump rate cannot exceed the rate of return of atoms to the ground state. If fresh atoms or molecules can be supplied to the mode volume, as in some gas and liquid lasers, then physical limitations on flow rates will control the pump rate. Thermal effects due to excessive optical pumping in solids can cause severe damage and thus limit maximum pumping levels. The fraction of power input that produces transitions to the pump band in gas lasers, frequently decreases with increasing current after some optimum current level is reached.

It is assumed that $N_1 = N_3 = 0$ for a four-level laser because of the fast transistion rates involved. The critical population inversion is once again given by $\Delta N_c = 1/(VB\tau_c)$ and at steady-state $\Delta N = \Delta N_c$. The critical pump rate W_{pc}, however, equals $\Delta N_c/(N_t\tau)$. In addition, because $\Delta N_c \ll N_t$, then clearly the critical pump rate for four-level lasers is generally much less than for three-level lasers—in fact, as much as 100 times less. (It is left as a problem for the reader to write down the rate equations for the four-level laser and deduce the results stated in this paragraph by analogy with the three-level laser development.)

6–9 OPTICAL RESONATORS

Some form of positive feedback of power is usually necessary if lasing action, in addition to gain, is to occur. This situation can be achieved by the use of distributed feedback, such as diffraction gratings, or by instantaneous feedback with mirrors. Distributed feedback is used with some diode lasers as well as in cases where lasers are integrated into a system, such as in fiber optics or integrated optical circuits. We will discuss only feedback in which mirrors are used. The fraction of power reflected by the output mirror varies from 98 or 99% for low-power lasers to less than 50% for some high-power systems.

Both stable and unstable resonators are used in lasers. A stable resonator is one in which, from a geometrical point of view, certain light rays retrace their paths indefinitely inside the resonator. Conversely, in an unstable resonator the light rays diverge or "walk off" the axis. Most laser resonator designs are stable, although unstable resonators have very useful properties for some high-power systems. Much more detail will be presented in Chapter 7 on the classification of laser resonators and their effects on laser beam properties. At this point, we will consider the various allowable modes of laser resonators.

If all laser resonators consisted of parallel-plane mirrors (Fabry-Perot interferometer), only axial (longitudinal) modes would need to be considered to determine allowed resonant frequencies. The existence of these modes is based on the electromagnetic boundary condition that only those modes will be supported by the resonator for which the phase change in one round trip, $2L$, is an integral multiple of 360° or, expressed mathematically,

$$m\lambda = 2L, \qquad m = 1, 2, 3, \ldots \qquad (6\text{--}26)$$

Be aware that L is the true optical path length of the laser. In Eq. (6–26) λ can be replaced by c/f. Then

$$f = mc/2L \qquad (6\text{--}27)$$

We see from Eq. (6–27) that the frequency spacing between adjacent axial modes is simply $c/2L$.

When one or more spherical mirrors are included in a stable resonator, the wavefronts are spherical, but the mirrors are not necessarily constant phase surfaces. In fact, any mode that is self-reproducing on one complete round trip is an allowable transverse mode. These modes are referred to as *transverse electromagnetic* (TEM_{lp}). They can be circular but frequently, especially if the laser output is not aperture limited, take a rectangular form. A few of these modes are depicted in Fig. 6–5.

The subscripts on TEM simply represent the number of nodes in orthogonal directions. Another important mode is the "doughnut" mode depicted in Fig. 6–6. Actually, this output is the result of a laser operating in the TEM_{10} (TEM_{01}) mode with rapid random variation of the orientation of the node in the plane perpendicular to the z axis (propagation axis).

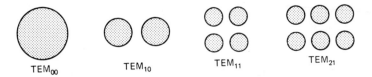

Figure 6-5   Transverse electromagnetic (TEM) modes.

The frequency of a given mode, including axial and transverse effects, can be deduced from electromagnetic theory and is given by

$$f_{nlp} = \left[n + (l + p + 1) \frac{\cos^{-1} \sqrt{g_1 g_2}}{\pi} \right] \left(\frac{c}{2L} \right) \qquad (6\text{-}28)$$

where $g_1 = 1 - L/r_1$, $g_2 = 1 - L/r_2$, and r_1 and r_2 are the radii of curvature of the resonator mirrors. Radii of curvature are taken as positive if measured toward the other mirror and negative otherwise. The frequency spacing between modes due to a difference of one in the transverse mode number is

$$\delta f = \frac{\cos^{-1} \sqrt{g_1 g_2}}{\pi} \left(\frac{c}{2L} \right) \qquad (6\text{-}29)$$

This spacing varies from zero to as much as c/L, which is, in fact, equivalent to $\delta f = 0$.

Unstable laser resonators must satisfy the same axial mode condition as stable resonators. Theoretically unstable systems can produce a variety of transverse modes; in practice, however, the losses associated with all but the lowest-order transverse mode are so high that unstable resonator lasers tend to operate in the lowest-order transverse mode. This mode is in the form of a ring whose dimensions depend on the resonator design. The ring output is due to geometrical walkoff of the beam, not diffraction, as is implied by the expression "diffraction coupling," which is often used to describe this type of resonator. A radially symmetrical substructure in the output ring is caused by diffraction. Diffraction effects also cause the output to be less than predicted geometrically. The ring

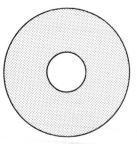

Figure 6-6   Doughnut mode. TEM_{10}^*

output of an unstable resonator should not be confused with the "doughnut" mode of stable systems, which occurs for an entirely different reason.

6-10 LINE BROADENING

Line broadening refers to the fact that some finite bandwidth must always be associated with any electromagnetic radiation regardless of its source. The radiation from a single isolated atom or molecule has a bandwidth due to what is referred to as natural broadening. Natural broadening is a result of the finite lifetime of an atom or molecule in an excited state. The average time that an atom spends in an excited state, prior to dropping to a lower state and giving off a photon, is called the lifetime of that state. According to a classical point of view (an incorrect point of view that yields essentially correct quantitative results in this case), the atom is a linear dipole that radiates for a period of time equal to the lifetime τ, thus producing a radiation pulse of length τ in time. It can be shown by Fourier analysis that a single pulse of time length τ must contain a continuum of frequencies with a full-width-at-half-maximum (FWHM) given approximately by

$$\Delta f \tau = 1 \qquad (6\text{--}30)$$

Thus the shorter the pulse, the greater is the bandwidth. The lifetime, τ, is typically 10^{-8} s. According to a quantum mechanical point of view (the correct point of view), Eq. (6–30) follows from the Heisenberg uncertainty relations.

Other broadening mechanisms become important in liquid, gas, or solid laser media. These mechanisms are referred to as homogeneous and inhomogeneous, terms that characterize the physical processes causing the broadening or, alternatively, as Lorentzian and Gaussian, respectively. The latter names characterize the shape of the lines (power versus frequency curves).

Homogeneous broadening is chiefly caused by collisions. In gaseous media collisions of the lasant atoms with each other, other species in the mixture, or with the walls confining the gas, will cause perturbations in the energy (frequency) of the photons emitted. By using the Heisenberg uncertainty principle, the frequency spread can be estimated in terms of the mean time between collisions t_c.

$$\Delta f t_c = 1 \qquad (6\text{--}31)$$

Frequent collisions result in a large bandwidth and so cooling a system to reduce thermal motion will decrease bandwidth. This phenomenon is referred to as homogeneous broadening because every atom in the gas is subject to the same probability of collision as every other atom. Each atom will have the same statistical distribution of frequencies of emission as all the others.

The mean time between collisions in a gas can be estimated by taking the ratio of the mean free path due to thermal motion, ℓ_{th} (average distance traveled between collisions) to the average thermal velocity, v_{th}. Equations for these quantities are readily obtained from thermodynamic arguments. Thus

$$\math{l}_{th} = \frac{2}{\pi} \frac{kT}{pa^2} \tag{6-32}$$

$$v_{th} = \left(\frac{kT}{\pi m} \right) \tag{6-33}$$

$$t_c = \frac{(mkT)}{16\pi pa^2} \tag{6-34}$$

where m is the molecular mass, p the gas pressure, k the Boltzmann constant, and a the molecular radius.

Homogeneous broadening in crystalline solids is the result of interactions between the lasant species and vibrations in the crystal lattice. Lattice vibrations are quantized and are referred to as *phonons*. So, in effect, interaction of lasant species with lattice vibrations is a particle-particle collision and can be treated just like collisions in a gas if the mean time between collisions, t_c, is known.

Inhomogeneous broadening occurs whenever any type of inhomogeneity causes some lasant species to differ from others in terms of the central frequency of their statistical distribution of emission frequencies. This situation occurs in solids as the result of any defects that make the environments of lasant species vary from point to point.

The most important form of inhomogeneous broadening in gas lasers is due to the Doppler effect. This effect refers to the fact that the frequency of light (as with sound) is a function of the relative speed of the source emitting the radiation and observer. The result of the application of the special theory of relativity to this phenomenon is shown as

$$f = \frac{(1 + v/c)f_0}{\sqrt{1 - v^2/c^2}} \tag{6-35}$$

where f_0 is the frequency measured by an observer at rest relative to the emitter, v is the relative speed, and c is the speed of light. The velocity is positive when the emitter and observer are moving toward each other and negative when they are moving away from each other. For lasers, $v \ll c$ so that Eq. (6-35) may be written

$$f = \left(1 + \frac{v}{c} \right) f_0 \tag{6-36}$$

The lasant atoms or molecules in a gas laser have, in general, components of their velocity in the direction of the axis of the laser. These velocity components are distributed according to Maxwell's velocity distribution

$$\frac{dN}{N} = \left(\frac{m}{2\pi kT} \right)^{-3/2} \left[\exp - \frac{m}{2\pi kT} (v_x^2 + v_y^2 + v_z^2) \right] dv_x \, dv_y \, dv_z \tag{6-37}$$

where dN/N is the fraction of atoms or molecules with velocity between **v** and **v** + **dv**, m is the mass of an atom, and k is the Boltzmann constant. If z is the axis of the laser, integration over the x and y velocity components yields

$$\left(\frac{dN}{N}\right)_z = \left(\frac{m}{2\pi kT}\right)^{1/2}\left[\exp\left(-\frac{m}{2\pi kT}\,v_z^2\right)\right]dv_z \tag{6-38}$$

Clearly the largest fraction of atoms per unit velocity component parallel to the z axis occurs for $v_z = 0$. As v_z increases, the fraction of atoms per unit velocity decreases rapidly [a plot of $N^{-1}(dN/dv)_z$ versus v_z yields a Gaussian curve]. The frequency emitted by the atoms is a linear function of their velocity, therefore the line shape will be the same as the velocity distribution—that is, the emitted power as a function of frequency will have a Gaussian shape centered around the frequency emitted for an atom at rest, f_0.

Doppler broadening is a form of inhomogeneous line broadening because each value of v_z represents a distinct group of atoms at any instant and these atoms can only emit at the frequency specified by Eq. (6–36). It can be shown that the Doppler bandwidth (FWHM) is given by

$$\Delta f = 7.16 \times 10^{-7}\sqrt{\frac{T}{M}}\,f \tag{6-39}$$

where T is absolute temperature and M is the mass in grams of one mole of gas.

Figure 6–7 illustrates both homogeneous and inhomogeneous line broadening. For comparative purposes, the bandwidth (FWHM) is assumed to be the same for both. The quantity $g(f)$ is called a *lineshape factor* and is defined such that $\int_0^\infty g(f)\,df = 1$; if I represents the total irradiance, then $g(f)I = I(f)$ gives the spectral irradiance.

It is now possible to modify Eq. (6–14) for the gain β to take into account line broadening effects. The gain must be proportional to the lineshape factor,

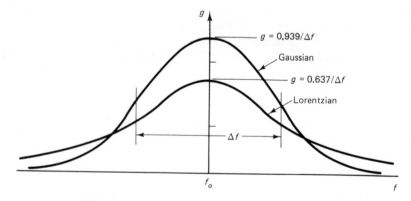

Figure 6–7 Gaussian and Lorentzian lineshapes compared.

for the more spontaneous power emitted at a given frequency, the greater is the gain at that frequency. Therefore the spectral gain is given by

$$\beta(f) = (N_2 - N_1) \frac{c^2 g(f)}{(8\pi f^2 \tau)} \tag{6-40}$$

Because the bandwidth is always small compared to f, the variation of f in Eq. (6-39) is negligible and $\beta(f)$ takes the same general shape as $g(f)$. It is worth noting that the maximum value of $g(f)$ is $0.637/\Delta f$ and $0.939/\Delta f$ for the Lorentzian and Gaussian lines, respectively [Svelto and Hanna 1976]. Thus the maximum gain as a function of population inversion can be estimated if the bandwidth and lifetime τ are known.

6-11 Q-SWITCHING, CAVITY DUMPING, AND MODE LOCKING

Techniques for modifying the laser output are discussed here, specifically, techniques for producing, short, high peak power pulses.

Q-switching.   This process involves interrupting the optical resonator in some fashion so that the population inversion process can progress with little or no feedback present to cause lasing. When the population inversion has reached some high level, the reasonator interruption is removed and the laser gives out a short pulse with very high peak power. This process is called *Q-switching* in reference to the *Q*-value of the resonator.

Q-value is defined by

$$Q = \frac{2\pi(\text{energy stored})}{\text{energy lost per cycle}} \tag{6-41}$$

If E is the energy stored in the cavity and τ_c is the lifetime of a photon in the cavity, Eq. (6-41) can be written

$$Q = 2\pi E \left(\frac{E}{f\tau_c} \right)^{-1} = 2\pi f\tau_c \tag{6-42}$$

The Q-value of any type of oscillating system can be shown to be given by the ratio of the oscillator central frequency to the oscillator bandwidth

$$Q = \frac{f}{\Delta f} \tag{6-43}$$

Hence

$$\Delta f = \frac{1}{2\pi \tau_c} \tag{6-44}$$

which would have been predicted on the basis of the uncertainty principle.[9] This is the ideal resonator linewidth and is much smaller than actual linewidths due to instabilities from vibrations and thermal fluctuations.

Q-value can also be related to cavity losses, particularly mirror reflectances. Assuming that R_1 and R_2 are the mirror reflectances and account for the only optical losses, Eq. (6–45) can be written

$$\frac{E}{\tau_c} = \frac{1 - R_1 R_2 E}{\tau_r} \qquad (6\text{–}45)$$

where τ_r is the round-trip time for the photon in the resonator. Equation (6–45) simply expresses the rate of energy loss in two different ways. Thus the average number of round trips for a photon is given by

$$\frac{\tau_c}{\tau_r} = \frac{1}{1 - R_1 R_2} \qquad (6\text{–}46)$$

If we use Eq. (6–46) and $\tau_r = 2L/c$, Eq. (6–42) then becomes

$$Q = \frac{4\pi f L}{c} (1 - R_1 R_2)^{-1} \qquad (6\text{–}47)$$

Equation (6–46) illustrates the fact that Q-values for lasers are quite high, typically of the order of a million.

Techniques for Q-switching have involved rotating mirrors and mechanical choppers for pulses on the order of 0.001 s. Acousto-optic modulators are used for pulses on the order of 10^{-6} s. Pockels and Kerr cells are used for pulses on the order of 10^{-9} s. The last three devices are discussed in more detail in Chapter 1.

Cavity dumping.   Certain types of lasers cannot be Q-switched successfully due to insufficient lifetime in the lasing state; then a procedure called cavity dumping may be used. This process literally allows the lasing energy to build up in the cavity with 100% feedback and then at some instant the feedback is reduced to a low value and a pulse of high energy is emitted. This technique does not yield the high peak power pulses of Q-switching but is used effectively, particularly in dye lasers.

Mode locking.   Lasers are generally capable of operating in a variety of modes, both axial and transverse. Normally when a laser is operating in several such modes at once, the individual modes (of different frequencies) are oscillating independently of each other and tend to have totally random phase relations relative to each other. Mode locking is a process in which large numbers

---

[9] Precisely stated, the uncertainty relation is $\Delta f \, \Delta t = 1/2\pi$.

of modes are placed in lock step with   result leads to very
short (as short as one picosecond or   high peak power.
     A common technique for ac   rough the use
of a saturable absorber. The abso.   anic dye,
that normally can absorb the radiation   modes
are in step, both in phase and in spatial   ply
more energy to the dye than it can absorb an   ge
of the pulse through the dye. All dye molecules are   ited
state. So the dye is saturated and briefly becomes t   nt, allowing part
of the pulse energy to pass through. The length of the cell containing the dye
is chosen to achieve exactly the right condition for saturation.
     Once such a mode-locked pulse occurs, part of it is naturally fed back
into the laser and with sufficient gain it saturates the dye each time it passes
through it. Thus once started, the pulse remains in the laser as long as pumping
continues. Some laser media are saturable and automatically produce mode
locking. Figure 6–8 schematically depicts the mode-locking process.

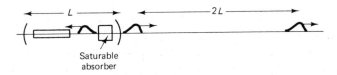

Saturable
absorber

Figure 6–8   Mode-locked pulses.

The distance between pulses is $2L$, the time between the pulses is $\tau_r =$
$2L/c$, and the bandwidth of the pulses is

$$\Delta f = N\,\delta f \tag{6–48}$$

where N is the number of modes locked and δf is the frequency spacing between them. According to the uncertainty principle, the temporal length of the
pulse is approximately $(N\,\delta f)^{-1}$. The spatial length (coherence length) is then
$c/(N\,\delta f)$. As many as 1000 modes may be locked together, therefore picosecond
length pulses can be obtained. Both pulsed and CW lasers can be mode locked.

PROBLEMS

6–1. (a) Calculate the ratio of the number of atoms in the excited state to the number
     in the ground state at $T = 300°K$, $700°K$, and $1000°K$ if the energy difference
     is 1 eV.
  (b) What are the wavelength and frequency of the light radiated in a transition
     from the excited state to the ground state?

6–2. Describe the three types of optical transitions that occur between energy states
     in a material.

6–3. According to Planck's theory, what is the radiation density of a blackbody for a frequency of 10^{14} Hz at $1000°$K?

6–4. Calculate the ratio of the spontaneous radiation coefficient to the stimulated radiation coefficient at a wavelength of 0.6328 μm.

6–5. If the round-trip gain in a 30-cm-long laser is 1%, calculate the net gain coefficient $(\beta-\alpha)$.

6–6. The spontaneous power output for a certain CW laser is 1 μW and the laser power output is 2 mW. If the length of the gain medium is 10 cm, calculate the gain coefficient. Assume that the only loss is due to 2% transmittance at the output mirror.

6–7. Discuss what is meant by three- and four-level lasing systems (illustrate with diagrams) and compare them.

6–8. If the lifetime of a lasing state is 0.5 ms, what is the approximate critical pump rate?

6–9. A solid laser consists of mirrors separated by 50 cm with a laser rod 10 cm in length with a refractive index of 1.5 and a Q-switch with a length of 2.5 cm and a refractive index of 1.45. What is the frequency spacing between axial modes for this laser?

6–10. Calculate the frequency spacing between a $TEM_{n\,10}$ and a $TEM_{n\,20}$ mode for a laser with an optical cavity length of 5 m and the following mirror radii:

$$r_1 = 30 \text{ m} \qquad r_2 = 20 \text{ m}$$

6–11. Estimate the mean time between collisions and the bandwidth due to collision broadening for a molecular gas (CO_2) at $400°$K and a pressure of 30 torr.

6–12. Show that the linewidth for a Doppler-broadened line can be written

$$\Delta\lambda = 7.16 \times 10^{-7} \sqrt{\frac{T}{M}} \, \lambda$$

where M is the mass in grams of a mole of gas.

6–13. Calculate the cavity Q-value for a CO_2 laser of 2-m cavity length and reflectances of 100 and 70%.

6–14. **(a)** What is the average number of round trips for a photon in a laser with mirror reflectances of 100 and 40%?

(b) If the optical cavity length is 1.5 m, what is the cavity lifetime?

6–15. A Nd-Glass laser ($\lambda = 1.06$ μm) has an optical cavity length of 1 m and 100 axial modes that are mode locked. What are the frequency bandwidth, spacing between pulses, time between pulses, coherence time, and coherence length of the pulses?

6–16. Calculate the Doppler frequency linewidth for a CO_2 laser if $T = 400°$K.

REFERENCE

SVELTO, O., and D. C. HANNA, *Principles of Lasers*. New York: Plenum Press, 1976.

BIBLIOGRAPHY

6–1 Verdeyen, J. T., *Laser Electronics.* Englewood Cliffs, NJ: Prentice-Hall, Inc., 1981.

6–2 Siegman, A. E., *An Introduction to Lasers and Masers.* New York: McGraw-Hill Book Co., 1971.

6–3 Maitland, A., and M. H. Dunn, *Laser Physics.* New York: John Wiley & Sons, Inc., 1969.

6–4 Young, M., *Optics and Lasers.* New York: Springer-Verlag, 1977.

6–5 Duley, W. W., *CO_2 Lasers: Effects and Applications.* New York: Academic Press, 1976.

chapter 7

Laser Optics

This chapter describes the propagation of laser beams and their focusing properties. In addition, the relationship of laser beam parameters, such as beam size and transverse modes, to optical resonator design is covered. Finally, both Gaussian (lowest-order TEM mode) beams and higher-order mode beams are discussed in connection with the aforementioned properties.

7–1 NATURE OF THE LASER OUTPUT

The beam emitted by a laser consists primarily of stimulated radiation with a small amount of spontaneous radiation (noise) included. Because the radiation is chiefly stimulated, a fairly high degree of temporal coherence occurs. Even in relatively inexpensive HeNe lasers a coherence length of 1 to 3 m is not unusual. Special stabilizing techniques can be used to increase the coherence length to 100 m fairly easily (Chapter 8). Spatial coherence across the wavefronts in gas lasers is nearly perfect. In solid lasers the transverse spatial coherence may be poor due to a tendency toward lasing in independent filaments within the crystal.

The output of some lasers is linearly polarized. In the case of HeNe lasers, this situation is the result of Brewster windows attached to the end of the discharge tube to contain the gas (Chapter 1). Brewster windows are positioned so that the axis of the laser is at the Brewster angle relative to the normal to the window as depicted in Fig. 7–1. Light reflected from any surface of one of the Brewster windows is polarized perpendicular to the plane of incidence

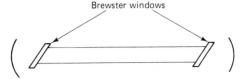

Figure 7-1   Gas laser with Brewster windows.

(plane of the paper in Fig. 7-1). None of the radiation polarized parallel to the plane of incidence is reflected. Therefore a large loss is introduced for the perpendicular component, resulting in a buildup of the parallel component. Nearly all the radiation in the resonator ends up with parallel polarization and so minimal (ideally zero) reflection losses occur at the windows. What reflection does occur is conveniently reflected *out* of the cavity.

Some crystalline lasers produce a polarized output because of optical anisotropy of the crystal. If ruby, which has a tetragonal crystal structure, is cut so that its *c* axis is perpendicular to the cavity a polarized output parallel to the *c* axis is produced. If the *c* axis is parallel to the laser axis, the output is randomly polarized.

A certain degree of linear polarization occurs in lasers where the beam path has been extended by folding the beam with mirrors if the angle between the beam and mirrors is near 45°. This result is clear for dielectric mirrors from the Fresnel equations presented in Chapter 1. Reflection for the perpendicular component is greater than for the parallel component (which is true even though the mirrors may be metal); as a result, this polarization state is preferentially amplified.

In all cases where linearly polarized output is not forced by polarizers, Brewster windows, special crystal properties, or folding mirrors, the output is instantaneously circularly polarized, but the phase fluctuates randomly with time. The fact that the radiation is stimulated results in instantaneous polarization. Thermal and mechanical instabilities cause timewise random changes in the phase of the polarization. Radiation with random polarization produces the same results as unpolarized radiation for most purposes.

It is extremely important to be aware of the nature of the polarization of the laser in many applications. When a laser beam must undergo many reflections in a plane, a large attenuation of the parallel component will occur. Consequently, almost no light is left if the output of the laser is linearly polarized in the parallel sense. More than one holographer has run afoul of this phenomenon. Linear polarization also causes a difference in cutting speed and quality for travel parallel and perpendicular to the polarization direction in high-power laser-cutting applications. Random and circular polarization exhibit no difference.

The beam from a laser expands as though emanating from a point source. In other words, the wavefronts for an axially symmetrical beam are spherical regardless of the TEM mode. The radius of curvature in the near field changes

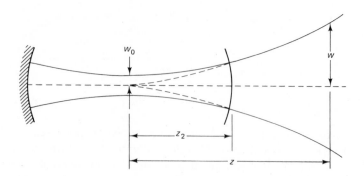

Figure 7-2 Propagating laser beam.

from an infinite value, at the waist, to a finite but linearly increasing quantity as the far field is approached. This situation is shown in Fig. 7-2.

As can be seen, the radius of curvature at the waist is infinite; as the far field is approached, the locus of points defining the beam size approaches an asymptote and from then on the beam appears to have originated from a point source located at the waist where the asymptotes intersect. As will be seen later, the far field is generally a *long* way from the laser and is rarely reached except in alignment, ranging, and surveying applications.

7-2 BEAM PROPAGATION
FOR STABLE LASER RESONATORS

The output of a stable laser resonator is usually a low-order transverse mode, frequently TEM_{00}, or some higher-order TEM_{lm} mode.[1] It is possible for the output to be multimode—that is, a mixture of various TEM modes. Several axial modes may be present, but we can ignore them as far as beam propagation is concerned.

Lowest-order mode beam propagation. Solution of the electromagnetic wave equation in rectangular coordinates leads to a particularly useful equation for describing beam propagation. Equation (7-1) gives the irradiance as a function of the cross-sectional coordinates x and y and the distance z from a point where the phase front is flat.

$$I(x, y, z) = I_0 \frac{\sigma_x(0)\sigma_y(0)}{\sigma_x(z)\sigma_y(z)} H_m^2 \left(\frac{x}{\sigma_x(z)}\right) H_n^2 \left(\frac{y}{\sigma_y(z)}\right)$$

$$\times \exp - \left[\frac{x^2}{\sigma_x(z)^2} + \frac{y^2}{\sigma_y(z)^2}\right] \quad (7-1)$$

[1] Resonator stability is discussed in Section 7-3.

The quantities σ_x and σ_y are proportional to the standard deviations of the Gaussian amplitude functions in the respective directions. I_0 is the irradiance at the center of the spot at $z = 0$, which is a point of minimum spot size, usually called the waist, and is normally (but not necessarily) located inside the cavity. H_m and H_n are Hermite polynomials, of which the first few are given here.

$$H_0(q) = 1 \qquad\qquad H_1(q) = 2q$$
$$H_2(q) = 4q^2 - 2 \qquad H_3(q) = 8q^3 - 12q \tag{7-2}$$

Two useful relationships for Hermite polynomials are presented for future reference.

$$H'_n = 2nH_{n-1}$$
$$H_{n+1} = 2qH_n - 2nH_{n-1} \tag{7-3}$$

Here the prime refers to differentiation with respect to the argument. The generating function for Hermite polynomials is given as Eq. (7–4).

$$H_n = (-1)^n e^{q^2} \frac{\partial^n}{\partial q^n} e^{-q^2} \tag{7-4}$$

Referring to Eq. (7–1), we see that the irradiance for the TEM_{00} mode at $z = 0$ is given by

$$I = I_0 \exp -\left(\frac{x^2}{\sigma_x^2} + \frac{y^2}{\sigma_y^2} \right) \tag{7-5}$$

This may be written

$$I = I_0 e^{-r^2/\sigma^2} \tag{7-6}$$

for a symmetrical beam.

The definition of spot radius may be taken to be σ, which represents the point at which the irradiance is one over e of its value at the center. A more practical definition of radius for TEM_{00} beams results when Eq. (7–6) is written

$$I = I_0 e^{-2r^2/w^2} \tag{7-7}$$

Then w is the point at which the irradiance has dropped to one over e^2 of its value at the center. This is the point at which the electric field amplitude has reached one over e of the central value. Over 86% of the total power is contained in a spot of diameter $2w$. Integration of the irradiance over a spot of diameter $2w$ (left as an exercise) yields the following relation between I_0 and P_t, the total power.

$$I_0 = \frac{2P_t}{\pi w^2} \tag{7-8}$$

The manner in which $w(z)$ varies with z has been determined by solving the scalar wave equation for a Gaussian beam and is presented here along with the equation for the phase front radius of curvature as Eq. (7–9) [Kogelnik and Li 1966].

$$w(z) = w_0 \left[1 + \left(\frac{\lambda z}{\pi w_0^2} \right)^2 \right]^{1/2}$$

$$R = z \left[1 + \left(\frac{z}{z_R} \right)^2 \right]$$

(7–9)

Here w_0 is the spot radius at $z = 0$ and $z_R = \pi w_0^2 / \lambda$ is called the Rayleigh range. Notice that $w(z)$ is symmetrical about $z = 0$. For very large values of z

$$\left(\frac{\lambda z}{\pi w_0^2} \right)^2 \gg 1$$

the divergence angle is given by $w(z)/z = \theta = \lambda / \pi w_0$. This is called the far-field divergence (diffraction) angle. It should be noted that this diffraction angle is 0.52 times the far-field diffraction angle of the central maximum for a plane wave passed through an aperture of diameter $2w_0$. Thus the fundamental (lowest-order Gaussian) beam propagates with about half the divergence of a plane wave of the same initial diameter in spite of the fact that the Gaussian beam is a spherical wave. Also, no higher-order diffraction maxima are associated with the Gaussian beam as with a plane wave incident on a finite aperture.

Equation (7–9) applies equally to a Gaussian laser beam emanating directly from the laser,[2] where z is then measured from the waist produced by the laser optical resonator, or to a focused laser beam. In the case of a focused beam, z is measured from the point of best focus where the phase front is planar. If the divergence angle of the beam entering a lens is small, the beam is focused approximately at the secondary focal point. Because the convergence is rapid (small focused spot size), the lens is in the far field relative to the focused spot. Therefore

$$w_0 = \frac{\lambda f}{\pi w_l}$$

(7–10)

where w_0 is the focused spot size and w_l is the spot size at the lens.

The depth of focus of a focused Gaussian beam may also be deduced from Eq. (7–9). Let $w(z) = \rho w_0$ at $z = \pm d$. Then

$$d = \pm \frac{\pi w_0^2}{\lambda} \sqrt{\rho^2 - 1}$$

(7–11)

---

[2] If the output mirror is not flat the lensing effect of the mirror should be taken into account.

where ρ is a measure of how much the spot radius has increased at a distance d from the focused spot. The depth of focus for a Gaussian beam is substantially greater than for a higher-order or plane wave of the same diameter. For $\rho = 1.05$, Eq. (7–11) reduces approximately to $d = \pm w_0^2/\lambda$.

Higher-order modes. Rectangular Hermite-Gaussian higher-order modes are discussed for two reasons. First, many lasers tend to operate in rectangular modes if they operate higher-order. Second, the essential features of all higher-order mode beam propagation are contained in this discussion and the mathematics are basically straightforward, most having been worked out long ago for the quantum-mechanical simple harmonic oscillator. Equation (7–1) is the basic description of the irradiance of a rectangular higher-order mode beam propagating in the z direction with x and y coordinate axes oriented parallel to the rectangular axes of the beam cross section.

William Carter proposed a definition of spot size for higher-order rectangular mode beams that uses the standard deviation of the Hermite-Gaussian functions [Carter 1980]. Thus the x and y axes are parallel to the sides of the rectangle defining the spot and the dimensions of the spot are $2\sigma_x(z)_m$ by $2\sigma_y(z)_1$.[3] Using results of the solution for a quantum-mechanical simple harmonic oscillator, for which Hermite-Gaussian functions give the probability function, Carter was able to show that

$$\sigma_s(z)_m = (2m + 1)^{1/2}\sigma_s(z) \qquad (7\text{–}12)$$

where $\sigma_s(z)_m$ is the spot size for a mode of order m, s is either the x or y direction, and $\sigma_s(z)$ is the spot size for the corresponding lowest-order mode. Figure 7–3 contains plots of $H_m^2 \ (s/\sigma_s) \exp \ (-s^2/\sigma_s^2)$ for several values of m. These plots represent the irradiance variations in either the x or y directions.

Notice in the cases shown in Fig. 7–3 that σ_{sm} falls outside the outermost irradiance peak. In fact, it may be argued that this is true for all values of m [Luxon and Parker 1981]. The expression $H_m^2 \exp \ (-s^2/\sigma_s^2)$ in quantum mechanics represents the probability (aside from a constant) of locating a particle undergoing simple harmonic motion at position s relative to its equilibrium position. The quantity σ_{sm} is shown to be the classical amplitude of the oscillator. As m (m is the quantum number) is increased, the correspondence principle of quantum mechanics predicts that the peak probability should occur at a value of s approaching σ_{sm}. At a value of $m = 14$ the peak probability has been calculated to occur at $s = 0.91\sigma_{sm}$. Under no circumstances can the peak

---

[3] $\sigma_x(z)_1$ is $\sqrt{2}$ times the standard deviation of the function

$$H_m^2 \left(\frac{x}{\sigma_x}\right) e^{(-x^2/\sigma_x^2)}$$

and similarly for $\sigma_y(z)_1$.

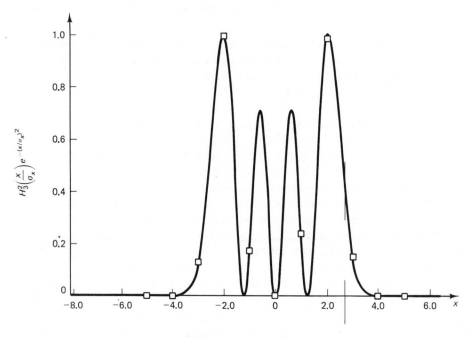

(a) Order number $m = 3$

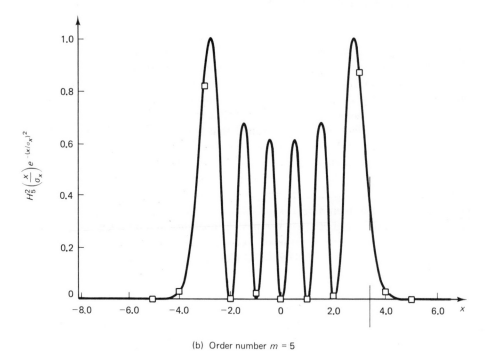

(b) Order number $m = 5$

Figure 7–3 Plots of $H_m^2(x/\sigma_x)e^{-(x/\sigma_x)^2}$ with $\sigma_x = 1$, maximum value $= 1$.

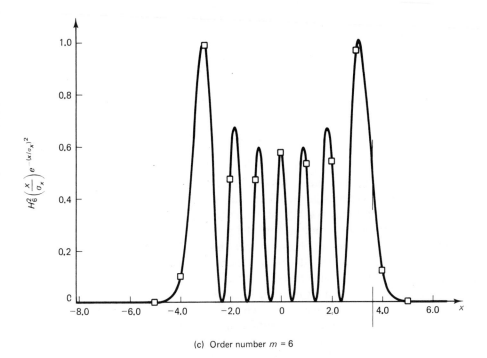

(c) Order number $m = 6$

Figure 7-3 (continued)

probability fall outside the classical amplitude. Therefore it may be concluded by analogy that all irradiance peaks for Hermite-Gaussian beams are contained within a spot of dimensions $2\sigma_{xl}$ by $2\sigma_{ym}$. Most energy in the beam falls within a spot of this size even for relatively low-order modes. Numerical integration for a TEM_{11} mode shows that nearly 80% of the beam energy falls within a spot size defined in this way; 85% of the energy falls within such a spot for a TEM_{44}.

It has been shown that the propagation of a higher-order rectangular beam may be written very simply in Eq. (7–9) for the lowest-order beam. This equation is repeated here with the beam size described in terms of σ, since the one over e^2 point has no significance for a higher-order beam.

$$\sigma_s(z) = \sigma_{s0}\left[1 + \left(\frac{\lambda z}{2\pi\sigma_{s0}^2}\right)^2\right]^{1/2} \qquad (7\text{–}13)$$

In Eq. (7–13) σ_s is the spot size for s equal to either x or y and σ_{s0} is the spot size for the focused beam. The modified equation for the spot size $\sigma_s(z)_m$ of the mth higher-order mode is

$$\sigma_s(z)_m = \sigma_{s0}(2m + 1)^{1/2}\left[1 + \left(\frac{\lambda z}{2\pi\sigma_{s0}^2}\right)^2\right]^{1/2} \qquad (7\text{–}14)$$

and $\sigma_{s0}(2m + 1)^{1/2} = \sigma_{s0m}$, which is the spot size for the focused beam. The far-field divergence angle is given by

$$\theta = \frac{\sigma_s(z)_m}{z} = \frac{(2m + 1)^{1/2}\lambda}{2\pi\sigma_{s0}} \tag{7-15}$$

It is instructive to express the far-field divergence angle in terms of the actual focused spot size, σ_{s0m}.

$$\theta = \frac{(2m + 1)\lambda}{2\pi\sigma_{s0m}} \tag{7-16}$$

Using Eq. (7–16), it is clear that the divergence angle for equal spot size is actually $(2m + 1)$ times larger for the higher-order beam.

The quantity σ_{sm} is a logical mathematical definition of spot size because all the irradiance peaks and most of the power fall within a spot of area $2\sigma_{xm}$ times $2\sigma_{yn}$. This definition, however, does not lend itself to direct measurement under any circumstances. It can be determined indirectly by making power measurements through apertures, a process with many difficulties. There is a more practical definition, at least for lasers where mode patterns can be readily viewed, such as burn patterns in acrylic for CO_2 lasers. This definition simply uses the distance between the centers of the outermost irradiance peaks in both the x and y directions. For convenience, half this distance will be used and symbolized by $D_s(z)_m$. Equation (7–14) is easily modified if the relation between $D_s(z)_m$ and $\sigma_s(z)_m$ is known. If we write

$$D_s(z)_m = \frac{1}{k_m}\sigma_s(z)_m \tag{7-17}$$

then Eq. (7–14) may be written

$$D_s(z)_m = D_{s0}(2m + 1)^{1/2}\left[1 + \left(\frac{\lambda z}{2\pi k_m^2 D_{s0}^2}\right)^2\right]^{1/2} \tag{7-18}$$

The values of k_m have been numerically computed for m values up to 14 and are listed in Table 7–1 [Luxon and Parker 1981].

Equation (7–18) is more useful if it is written entirely in terms of the focused spot size $D_{s0m} = D_{s0}(2m + 1)^{1/2}$.

$$D_s(z)_m = D_{s0m}\left[1 + \left(\frac{\lambda z(2m + 1)}{2\pi k_m^2 D_{s0m}^2}\right)^2\right]^{1/2} \tag{7-19}$$

The far-field divergence angle may now be written

$$\theta = \frac{\lambda(2m + 1)}{2\pi k_m^2 D_{s0m}} \tag{7-20}$$

When a laser beam is focused with a mirror or lens, the convergence is so rapid that the focusing element is in the far field in terms of distance from

TABLE 7-1   Values of k_m for
$m = 1$ through $m = 14$

m	k_m
1	1.73
2	1.42
3	1.30
4	1.24
5	1.20
6	1.18
7	1.16
8	1.14
9	1.13
10	1.12
11	1.11
12	1.11
13	1.10
14	1.10

the focused spot. Assuming that the best focus occurs, to a sufficient approximation, at the focal point, Eq. (7–19) reduces to Eq. (7–21) for the focused spot size.

$$D_{s\,om} = \frac{\lambda f(2m + 1)}{2\pi k_m^2 D_s(f)_m} \tag{7-21}$$

It is worth noting the inverse relationship between spot size and divergence angle. This leads to the following important result: when a laser beam of any order is expanded and recollimated, the divergence angle decreases proportional to the increase in beam size.

A useful definition of depth of focus for higher-order mode beams can be deduced from Eq. (7–19) in the same manner that Eq. (7–9) was deduced from Eq. (7–8). Again, let $z = d$, $D_s(d)_m = \rho D_{s\,om}$, and solve for d:

$$d = \pm 2\pi(\rho^2 - 1)^{1/2}\frac{k_m^2 D_{s\,om}^2}{(2m + 1)\lambda} \tag{7-22}$$

It is important to note the inverse effect of the mode number on depth of focus. This effect is usually more than compensated for, however, by the larger focused spot size associated with high-order beams.

The higher-order mode equations can be applied to circular (Laguerre-Gaussian) modes by setting $k_l = 1$ and replacing $(2l + 1)$ by $2(2p + m + 1)$, where p and m are the number of angular and radial nodes, respectively [Phillips and Andrews 1983]. The beam radius, in this case, relates to the fraction of power that will pass through an aperture of this radius centered on the beam axis.

7–3 RESONATOR STABILITY AND BEAM PARAMETERS

When referring to the optical stability of a laser resonator, we are talking about whether any rays, geometrically speaking, are trapped within the resonator. We can model the optical cavity as an infinite series of lenses with alternating focal lengths equal to the focal lengths of the mirrors with spacing between them equal to the optical length of the cavity. If any rays propagating near the axis of such an infinite series of lenses are periodically refocused, the system is stable. If they all diverge, it is unstable. By a purely geometrical analysis, it can be shown that all stable cavities obey the following relation.

$$0 \le g_1 g_2 \le 1 \tag{7-23}$$

Here $g_1 = 1 - L/r_1$ and $g_2 = 1 - L/r_2$, where r_1 and r_2 are the radii of curvature of the back mirror and the output mirror, respectively, and L is the resonator optical length. These g factors were introduced earlier while discussing the frequencies of TEM modes. A stability diagram is constructed by plotting $g_1 g_2 = 1$ in Fig. 7–4. Points lying on the lines or in the shaded region correspond to stable laser resonators. Points lying on the axes or the hyperbola are referred to as marginally stable because a slight shift in laser length or mirror radius of curvature can cause the system to become unstable.

Some points of particular interest on the stability diagram are

1. $g_1 = 1$, $g_2 = 1$, which represents the plane parallel mirror Fabry-Perot interferometer.

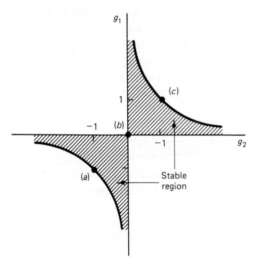

(a) Concentric symmetrical
(b) Confocal symmetrical
(c) Plane mirrors

Figure 7–4 Stability diagram for laser resonators.

2. $g_1 = 0$, $g_2 = 0$, which is a symmetrical confocal resonator. This configuration leads to the lowest diffraction losses in the TEM$_{00}$ mode compared with the higher TEM modes.

3. $g_1 = -1$, $g_2 = -1$, which is a symmetrical concentric configuration.

A simple rule to determine whether a resonator is stable can be stated. *If either the center of curvature of one mirror or the mirror itself, but not both, falls between the other mirror and its center of curvature, then the reasonator is stable* [Yariv 1968]. If this rule is violated, the resonator is unstable.

Many factors affect the selection of a particular mirror configuration. One mentioned earlier is whether TEM$_{00}$ operation is desired. The confocal resonator exhibits the lowest diffraction losses and the highest selectivity due to a large differential in diffraction losses for the TEM$_{00}$ mode and higher-order modes. This configuration also does a fair job in regard to maximizing mode volume. Mode volume is the volume of the active medium that is actually filled by radiation. Several resonator mirror configurations are shown in Fig. 7–5, along with the approximate mode volume for each.

The confocal mirror configuration is desirable from the standpoint of TEM$_{00}$ mode selectivity, but it is quite hard to align because x, y and angular alignment of the mirrors are critical. The concentric configuration has the same problem. Both are marginally stable. The problem of marginal stability is not too difficult because the advantage of the confocal resonator can be nearly achieved by selecting a configuration that lies safely in the stable region of Fig. 7–4 but that is close to the origin.

The plane mirror arrangement has the best mode-volume-filling capability but results in an output that is heavily multimode and it is a difficult system to align. Furthermore, the TEM modes are degenerate, making low-order transverse mode operation impossible.

A popular configuration is part (d) in Fig. 7–5 because of the relative ease of alignment and good mode-filling characteristics. This arrangement is frequently used in high-power lasers, with the radius r_1 being several times the optical length of the laser.

Whether a stable resonator laser operates in the lowest-order mode or higher-order largely depends on the size of the effective aperture in relation to the cavity length. A gas laser with a long narrow bore tubular arrangement will tend to operate low order because of the attenuation of the higher-order

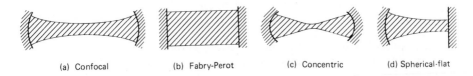

(a) Confocal   (b) Fabry-Perot   (c) Concentric   (d) Spherical-flat

Figure 7–5   Stable resonator configurations.

modes by the inner walls of the tube. Remember, the higher-order modes are associated with higher diffraction or divergence angles and so are more strongly attenuated by a limiting aperture.

A measure of the tendency for a stable laser cavity to operate low or higher order is the Fresnel number, given by

$$N = \frac{a^2}{\lambda L} \qquad (7\text{–}24)$$

where a, the effective aperture radius, should be taken as the radius of the smallest aperture in the system if it is not too far from one of the mirrors. In a system where the only limitation is due to the mirrors, assuming that they have equal radii (otherwise let $a^2 = a_1 a_2$, where a_1 and a_2 are the radii of the two cavity mirrors), the Fresnel number represents the number of Fresnel zones (interference rings) intercepted by one of the mirrors if a uniform plane wave illuminates the other mirror. A Fresnel number of one or less generally results in lowest-order mode (TEM$_{00}$) operation.

Stable cavity parameters.   Expressions for the waist size and its location relative to the laser mirrors have been derived for lowest-order Gaussian and higher-order mode cases [Kogelnik and Li 1966]. For the lowest-order Gaussian, the waist radius formed inside (or outside) the laser cavity is given by

$$w_0 = \left(\frac{\lambda L}{\pi}\right)^{1/2} \frac{[g_1 g_2 (1 - g_1 g_2)]^{1/4}}{(g_1 + g_2 - 2g_1 g_2)^{1/2}} \qquad (7\text{–}25)$$

where w_0 is the radius measured to the one over e squared irradiance points. Notice that Eq. (7–25) is an indeterminate expression for certain values of g_1 and g_2. When $g_1 = g_2 = 1$ (two plane mirrors), the spot size is aperture limited and has nothing to do with the resonator parameters. For the $g_1 = g_2 = 0$ case (symmetrical confocal resonator), if g_1 and g_2 are replaced by g and some rearranging carried out, Eq. (7–25) reduces to

$$w_0 = \left(\frac{\lambda L}{2\pi}\right)^{1/2} \qquad (7\text{–}26)$$

The location of the beam waist relative to the output mirror, determined by purely geometrical means, is given by

$$z_2 = \frac{g_1 (1 - g_2) L}{g_1 + g_2 - 2g_1 g_2} \qquad (7\text{–}27)$$

Waist location has no meaning for the Fabry-Perot arrangement. The waist is located midway between the mirrors for all symmetrical resonators. For lasers with a plane output mirror and spherical back mirror, the waist is located at $z_2 = 0$ or right at the output mirror.

Based on Carter's results, Eq. (7–23) has been modified by Luxon and Parker for Hermite-Gaussian higher-order modes [Luxon and Parker 1981]. By using the definition of spot size introduced earlier for higher-order modes (half the distance between outer irradiance peaks), the waist size becomes

$$D_s(0)_m = \left[\frac{\lambda L(2m+1)}{2\pi k_m^2}\right]^{1/2} \frac{[g_1 g_2 (1 - g_1 g_2)]^{1/4}}{(g_1 + g_2 - 2g_1 g_2)^{1/2}} \qquad (7\text{–}28)$$

The relation for the location of the waist is not altered by higher-order mode operation, for this location depends on purely geometrical factors.

Unstable resonators.   Unstable optical resonators are frequently used in high-power, high-gain lasers in which a relatively low percentage of feedback is required to achieve lasing operation. Although an infinite number of possible unstable resonator configurations are possible, Fig. 7–6 typifies the type used in commercial lasers, particularly multikilowatt CO_2 lasers.

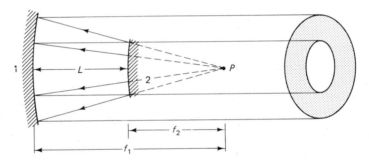

Figure 7–6   Unstable confocal resonator.

In an unstable cavity the rays "walk off" the axis and exit past the output mirror unless some mirror arrangement is used to intercept the rays and deflect them in some other direction (as is usually the case). In any event, the "walk off" of the rays leads to a ring-shaped output in the near field with some substructure due to diffraction effects.

Unstable resonators are capable of producing a variety of transverse modes, just as stable resonators do. Yet the losses associated with the higher-order modes are so great that, in practice, such lasers nearly always operate in their lowest-order transverse mode. This mode can result in a divergence angle and focusing capability approximately equal to the Gaussian mode. When focused, the irradiance distribution is similar to that of a focused Gaussian beam. The hole in the center of the beam in the far field is filled in and the irradiance distribution again is approximately Gaussian.

Some useful results can be deduced from Fig. 7–6 for a confocal unstable resonator. The ratio of a_1 to a_2, the outer and inner radii of the output ring, is called the magnification m and is related by similar triangles to the mirror radii of curvatures by

$$m = \frac{a_1}{a_2} = \frac{r_1/2}{r_2/2} = \frac{r_1}{r_2} \tag{7-29}$$

where use has been made of the fact that the focal length of a spherical mirror is half its radius of curvature. Note that a_2 is the radius of the output mirror but a_1 is the minimum radius of the back mirror.

If it is assumed that the radiation uniformly fills the back mirror up to a radius a_1, the fraction of the power emitted (geometrical loss G) is given by

$$G = \frac{\pi(a_1^2 - a_2^2)}{\pi a_1^2} = \frac{m^2 - 1}{m^2} = \frac{r_1^2 - r_2^2}{r_1^2} = 1 - \frac{r_2^2}{r_1^2} \tag{7-30}$$

This provides a reasonable estimate of the radiation geometrically coupled out of a confocal unstable resonator. It is also useful to note that

$$L = f_1 - f_2 = \frac{r_1}{2} - \frac{r_2}{2} \tag{7-31}$$

where all quantities are taken to be positive.

An equivalent Fresnel number can be defined for unstable resonators and for the confocal resonator discussed here [Svelto 1976].

$$N_{eq} = \frac{1}{2} \frac{a_2^2}{\lambda L} (m - 1) \tag{7-32}$$

For m near 1 $N_{eq} < N$ and excellent mode discrimination occurs for all half-integer values of N_{eq}.

Effective thin lens equation.   S. A. Self has developed a useful equation that resembles the thin lens equation which relates Gaussian waist locations to lens focal length and the Rayleigh range [Self 1983]. A thin lens transforms the radius of curvature of a Gaussian beam according to

$$\frac{1}{R_1} + \frac{1}{R_2} = \frac{1}{f} \tag{7-33}$$

where R_1 is radius of curvature entering the lens and R_2 is radius of curvature leaving the lens. The sign convention is the same as of o, i, and f for a thin lens as given in Chapter 1. Using Eq. (7-33) along with Eqs. (7-9) and the fact that a thin lens leaves the beam size unchanged immediately on either side of the lens leads to

$$\frac{1}{o + Z_R^2/(o - f)} + \frac{1}{i} = \frac{1}{f} \tag{7-34}$$

where o is the waist location for light entering the lens and i is the waist location for light leaving the lens. The sign convention on o and i is still the same as in Chapter 1.

Self also presented an expression for magnification

$$m = [(1 - o/f)^2 + (Z_R/f)^2]^{-1/2} \qquad (7\text{-}35)$$

which can be used for spot size determination

It has been pointed out by Luxon et. al. 1984, that Eqs. (7–34) and (7–35) are valid for beams of all modes. For Hermite-Gaussian beams the Rayleigh range is given by Eq. (7–36)

$$Z_R = \frac{2\pi k_m^2 D_{s0}^2}{\lambda(2m + 1)} \qquad (7\text{-}36)$$

It should be noted that Z_R is purely geometrical and can be calculated from

$$Z_R = L \frac{[g_1 g_2(1 - g_1 g_2)]^{1/2}}{g_1 + g_2 - 2g_1 g_2} \qquad (7\text{-}37)$$

if the resonator parameters are known.

PROBLEMS

7–1. The total power emitted by a CW laser operating in the TEM_{00} mode is 10 W. If the focused spot size is 0.2 mm, calculate the irradiance at the spot center.

7–2. The beam radius for a HeNe laser ($\lambda = 0.6328~\mu m$) operating TEM_{00} is 0.5 mm. Determine the spot size at distances of 1 m and 500 m from the waist.

7–3. If the spot radius for a HeNe TEM_{00} beam entering a 12.5-cm focal length lens is 1.5 cm, calculate the focused spot size.

7–4. Calculate the depth of focus of the conditions given in Problem 7–3 for a spot size increase of 10%.

7–5. A CO_2 laser ($\lambda = 10.6~\mu m$) operates in a rectangular TEM_{64} mode. The beam dimensions, outer peak-to-peak distance, are 2.5 and 2.3 cm. Calculate the focused spot size for both spot dimensions for focusing with a 25-cm focal length lens.

7–6. Calculate the depth of focus for the conditions given in Problem 7–5 for a 10% spot size increase.

7–7. A HeNe laser operating TEM_{00} has a resonator length of 1 m and mirror radii of $r_1 = 3$ m and $r_2 = 2$ m. Locate the waist and determine the spot size at the waist. Also, determine the spot size and divergence angle at the output mirror (r_2).

7–8. A CO_2 laser operating rectangular TEM_{57} has a resonator length of 5 m and mirror radii $r_1 = 30$ m, $r_2 = 20$ m. Locate the waist and determine spot dimensions at the waist. Also, determine the spot dimensions and divergence at the output mirror (r_2).

7-9. Prove whether the following mirror configurations are stable and plot them on a stability diagram:
   (a) $L = 2$ m, $r_1 = r_2 = 2$ m
   (b) $L = 1.5$ m, $r_1 = r_2 = 0.75$ m
   (c) $L = 5$ m, $r_1 = 30$ cm, $r_2 = 20$ m
   (d) $L = 1$ m, $r_1 = 4$ m, $r_2 = -2$ m
   (e) $L = 1.5$ m, $r_1 = 3$ m, $r_2 = -2$ m
   (f) $L = 3$ m, $r_1 = 2$ m, $r_2 = 2.5$ m

7-10. Design a stable laser resonator of 30-cm length that will provide a waist at the output mirror with a diameter of 1 mm for a HeNe laser operating TEM_{00}. What should the maximum aperture be for this laser to ensure TEM_{00} operation?

7-11. Design a confocal unstable resonator with a length of 1 m and an output mirror radius of 2.5 cm that will provide 40% output coupling. What are the magnification and equivalent Fresnel number for this resonator? Do you expect good mode discrimination for this resonator if $\lambda = 10.6$ μm?

7-12. For Problem 7-8 treat the output mirror as a thin lens and determine the new waist location, size and Rayleigh range. (Assume $n = 2.4$ and the outer surface is flat.) Use this information to locate the waist and its size produced by a 12.7 cm focal length lens placed 10 m from the output mirror. Compare your results with the results based on the assumption that the waist produced by the lens occurs in the focal plane.

REFERENCES

CARTER, W. H., "Spot Size and Divergence for Hermite-Gaussian Beams of Any Order," *Applied Optics*, **19**, No. 7, April 1980, 1027–1029.

KOGELNIK, H., and T. LI, "Laser Beams and Resonators," *Applied Optics*, **5**, No. 10, October 1966, 1550–1567.

LUXON, J. T., and D. E. PARKER, "Higher-Order CO_2 Laser Beam Spot Size and Depth of Focus Determination," *Applied Optics*, **21**, No. 11, June 1, 1981, 1933–1935.

PHILLIPS, R. L., and L. C. ANDREWS, "Spot Size and Divergence for Laguerre Gaussian Beams of any order," *Applied Optics*, **22**, No. 5, Mar. 1, 1983, 643–644.

SELF, S. A., "Focusing of Spherical Gaussian Beams," *Applied Optics*, **22**, No. 5, March 1, 1983, 658–661.

SVELTO, O., *Principles of Lasers*. New York: Plenum Press, 1976, 135–142.

YARIV, A., *Quantum Electronics*. New York: John Wiley & Sons, Inc., 1968, 226.

BIBLIOGRAPHY

7-1. Verdeyen, J. T., *Laser Electronics*. Englewood Cliffs, NJ: Prentice-Hall, Inc., 1981.

7-2. Young, M., *Optics and Lasers: An Engineering Approach*. New York: Springer-Verlag, 1977.

7-3. Siegman, A. E., *An Introduction to Lasers and Masers*. New York: McGraw-Hill Book Co., 1971.

7–4. Maitland, A., and M. H. Dunn, *Laser Physics*. New York: John Wiley & Sons, Inc., 1969.

7–5. Nemoto, S., and T. Makimoto, "Generalized Spot Size for a Higher-Order Beam Mode," *J. Opt. Soc. Am.*, **69**, No. 4, April 1979, 578–580.

7–6. Luxon, J. T., and D. E. Parker, "Practical Spot Size Definition for Single Higher-Order Rectangular-Mode Laser Beams," *Applied Optics*, **21**, No. 11, May 15, 1981, 1728–1729.

7–7. Luxon, J. T., D. E. Parker, and J. Karkheck, "Waist Location and Rayleigh Range for Higher-Order Mode Laser Beams," submitted to *Applied Optics*, July 1984.

chapter 8

Types
of Lasers

Most of the lasers found in industrial applications are discussed in this chapter, including some, such as the dye laser, that are used primarily in laboratory facilities. In most cases, only general descriptions of wavelength, output power, mode of operation, and applications are given. As a specific example, the CO_2 laser is discussed in some detail.

Laser types can be categorized in a number of ways. In general, there are gas, liquid, and solid lasers. Subclassifications of neutral gas, ion, metal vapor, and molecular lasers exist in the category of gas lasers. The only liquid laser discussed here is the dye laser. Solid lasers may be crystalline, amorphous (at least in the case of the Nd-Glass laser), or even a semiconductor.

8-1 HeNe LASERS

The HeNe laser is the most common of all the visible output lasers (see Fig. 8-1 for a photograph). It emits in the red part of the visible spectrum at 0.6328 μm. The power levels available range from 0.5-mW to 50-mW CW output. Most HeNe lasers are constructed with long, narrow bore (about 2 mm diameter is optimum) tubes, resulting in preferential operation in the TEM_{00} mode, although TEM_{10}^* may occur if good alignment of the mirrors is not maintained.

The electrical glow discharge is produced by a dc voltage of 1 to 2 kV at about 50 mA. A starting voltage of several kilovolts is applied to get the discharge started. The HeNe laser is a low-gain device that requires optimally

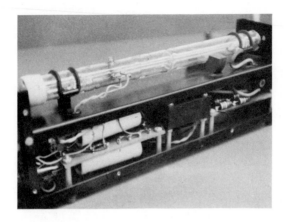

Figure 8–1   Photograph of a 1-mW HeNe laser with the cover removed.

99% feedback to sustain oscillation. Consequently, the mirrors are multilayer dielectric on a dielectric substrate for both back and output mirrors. The output power is 25 mW/m of tube length.

HeNe lasers have a gas mixture with Ne at a partial pressure of 0.1 torr[1] and a total pressure of 1 torr. Ne is the lasant, but He plays an important role in the excitation process. In a four-level process He is excited by collisions with electrons to a level coincident with an excited state of Ne. During collisions, the energy is resonantly transferred from He to Ne. The Ne decays to one of three possible intermediate states with three possible emission lines, one at 0.6328 μm and two in the ir at 1.15 and 3.39 μm. The mirrors act as selective filters for the 0.6328-μm wavelength.

Figure 8–2 shows the energy level diagrams of He and Ne and the pertinent transitions. The excited states of He that are involved are metastable, thus ensuring a high probability of resonant energy transfer to Ne. Neon atoms relax to the 1 S level after lasing. This state is metastable and must be depopulated by collisions with the wall of the tube. Therefore increasing the bore of the tube past about 2 mm will decrease power output rather than increase it.

Some HeNe lasers are designed to operate in a few low-order modes, so-called multimode operation. They may also be purchased with either linear or random polarization. The mirrors are attached directly to the tube for random polarization. At least one Brewster window is used for linearly polarized output.

The bandwidth of HeNe lasers is 1500 MHz because of Doppler broadening. Consequently, these lasers generally operate in several axial modes. Shortening the cavity to about 10 cm usually ensures single axial mode operation.

In single axial mode operation there will be just one dip in the gain curve due to "hole burning." This dip can be centered by adjusting the cavity length.

---

[1] A torr is approximately 1 mm of mercury.

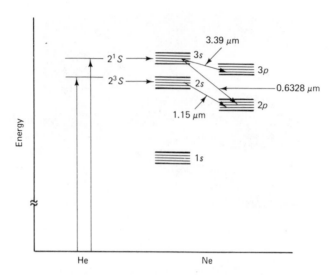

Figure 8–2   Energy level diagram for HeNe laser.

As the laser frequency drifts from this central value, the power will begin to increase (a transient effect). The power change can be monitored by a beam splitter and photodetector. If one mirror is mounted on a piezoelectric crystal, a servosystem can be used to stretch or contract the crystal slightly (by means of an applied voltage), thus adjusting the cavity length continuously to maintain minimum power. In this manner, a frequency stability of one part in 10^9 can be achieved.

High stability can be gained by placing a gas in a separate cell with a coincident absorption line in the laser cavity. I_2^{129} is used with HeNe lasers. I_2^{129} has a very narrow absorption line at 0.633 μm. By locking the HeNe laser on the center of this absorption and saturating the I_2^{129}, frequency stability of one part in 10^{12} to 10^{13} can be achieved.

The diversity of HeNe laser applications is too great to cover exhaustively and so some representative applications are mentioned here. More detail on selected low-power applications can be found in Chapters 9 and 10.

HeNe lasers are used in bar code reading, image and pattern recognition, alignment operations in industry, construction, sewer, and tunnel work, and surveying. In addition, they have a variety of nondestructive testing (NDT) applications, including surface flaw and roughness determination. Measurement applications include interferometric displacement measurement, absolute size measurement, and edge location, using various types of detectors and detector arrays. Frequently HeNe lasers are used to align other lasers and to locate the point of incidence on a work piece for a high-power laser.

HeNe lasers are also used in line-of-sight communications applications, such as control of construction equipment like earth graders to maintain proper grade or depth requirements. A major application of HeNe lasers is in the recording and playing back of holograms.

8-2 ION LASERS

Common ion lasers are argon (Ar), krypton (Kr), which has nothing to do with the Superman, and xenon (Xe). The power levels of these lasers vary from milliwatts to several watts. The major wavelengths available for each are listed in Table 8-1.

TABLE 8-1   Wavelengths and power ranges
for ion lasers

Laser	Wavelength (μm)	Power Range (W)
Ar	0.5145 ⎱ 0.488 ⎰	0.005–20
Kr	0.6471	0.005–6
Xe	0.995–0.5395	200 pulsed

The ion lasers, as the name implies, use an ionized gas as the lasant in an electrical discharge in a plasma. Basically they are four-level systems, but it is the excited states of the ions that are involved in the lasing process. Excitation is a two-collision process: the first collision ionizes the atom; the second provides the necessary excitation.

The current densities required in ion lasers are about 1000 A/cm², which limits the tube bore diameter to a few millimeters. The ions drift toward the cathode and so a return path outside the discharge is provided, usually by means of a segmented BeO tube. An axial magnetic field is applied by means of a current-carrying coil wrapped around the tube. This step tends to prevent electrons from losing energy through collisions with the wall since the magnetic field causes them to undergo a spiral motion along the tube axis.

Argon lasers are used extensively in surgery, particularly of the eye. This device is used to bounce light off the retroreflectors that the astronauts placed on the moon for moonquake and meteor impact studies as well as measurement of the distance between the earth and moon. The Ar laser, along with the Kr laser, is used in spectroscopic work, such as Raman spectroscopy. Both lasers are tunable over several wavelengths in addition to the ones listed in Table 8-1. The Xe laser can be used for some materials processing, such as cutting or removal of thin metallic films. In this type of work the Xe laser is operated in pulsed fashion with peak powers of around 200 W.

The Ar laser has been used in holographic work to achieve shorter exposure times and thereby minimize the effects of vibration. The development of high-speed, high-resolution emulsions for holography has reduced the need for Ar lasers in holographic work.

8–3 METAL VAPOR LASERS

Many types of metal vapor lasers exist, but the He-Cd and He-Se lasers are the most important commercial ones. Although essentially ion lasers, they behave similarly to the HeNe laser in that the metal ions are excited by collisions with excited He atoms. The excitation process in the He-Cd laser is called *Penning ionization*. In this process, an excited metastable He atom collides with a Cd atom, thereby ionizing it and raising it to an excited ionic state. Lasing action takes place between the excited ionic state and the ionic ground state of Cd. The He-Cd laser has wavelengths at 0.441 μm in the blue and 0.325 μm in the uv. Excitation in the He-Se laser is by charge transfer. An excited ionized metastable He atom collides with a neutral Se atom, transfers the charge to it, and excites it. Emission occurs between excited states of the Se ion with about 19 lines possible in the visible spectrum.

In metal-vapor lasers the metal is vaporized near the anode and condenses at the cathode, a process referred to as cataphoresis. Consequently, the metal is consumed, for example, at a rate of 1 g of Cd per 1000 hours of operation.

The power output of these metal-vapor lasers is in the 50- to 100-mW range. The current densities required for the one-step excitation process in metal-vapor lasers are much lower than for single-element ion lasers and so they can be air cooled and operated from 110-V ac.

Some applications for metal-vapor laser are light shows, full color image generation, spectroscopy, and photochemical studies.

8–4 DYE LASERS

Dye lasers, as the name suggests, use an organic dye as the lasing medium. Commercial dye lasers are pumped by nitrogen, argon, Nd-YAG, and excimer lasers, although Xe lamps can be used. Operation is usually pulsed at 100 pps, but average powers of 10 mW are attained.

The organic dyes used, such as rhodamine 6G (xanthene dye), have hundreds of overlapping spectral lines at which they can lase. Consequently, a dye laser with a specific dye can, in effect, be tuned continuously over a large portion of the visible part of the spectrum. By using several dyes, which are automatically interchanged, the entire visible portion of the spectrum, as well as some of the ir and uv, can be scanned with relatively equal power output at each wavelength. This capability is of great value in certain types of spectroscopic investigations in or near the visible wavelengths and also in photochemistry, including uranium isotope separation, and pollution detection.

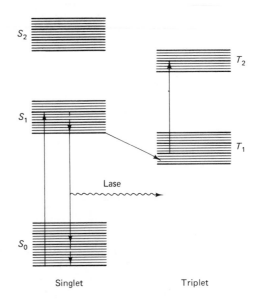

Figure 8–3 Energy levels for an organic dye.

Singlet   Triplet

Figure 8–3 is a representative energy level diagram for an organic dye. The diagram is composed of two types of states: the singlet states labled S_0, S_1, and S_2 and the triplet states labeled T_1 and T_2.[2] Each band of energies consists of widely spaced vibrational levels and overlapping rotational levels. Radiative transitions are quantum mechanically forbidden between S and T levels, but transitions between one S level and another or one T level and another are allowed. The dye molecule is optically pumped to the S_1 level and drops to the lowest level in that band. Lasing takes place between that level and one of the many possible S_0 states. Continuous operation is not practical in most systems because intersystem crossing from S_1 to T_1 states, caused by collisions, results in a buildup of dye molecules in the T_1 states and optical absorption from T_1 to T_2 states. Pulse lengths are kept short compared with the intersystem crossing time of about 100 ns.

The bandwidth of dye lasers is 10^{13} Hz, which makes it possible to achieve picosecond pulses by mode locking.

8–5 RUBY LASER

The ruby laser was the first optical maser (or laser, as it came to be called) built. The first such laser was constructed by Maiman at the Hughes Laboratories, Malibu, California, in 1960. Figure 8–4 contains a schematic representation of a ruby laser.

The active material in the ruby laser is triply ionized chromium, Cr^{3+},

---

[2] Singlet and triplet states have zero and one, respectively, for the total spin of electrons in the molecule and determine the multiplicity of the energy levels.

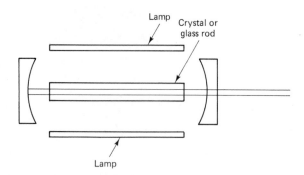

Figure 8–4 Schematic of solid laser—for example, ruby, Nd-Yag, Nd-Glass.

and the host is crystalline aluminum oxide (sapphire), Al_2O_3. Cr_2O_3 substitutes for about 0.05% by weight of the Al_2O_3, providing the crystal with a characteristic light pink color. The deep red color of gemstone ruby is the result of much higher concentrations of Cr_2O_3. The output of the ruby laser is at 0.6943 μm in the red.

Ruby lasers are operated in pulsed fashion with repetition rates the order of one pulse per second. The energy per pulse may be as high as 100 J but is usually a few joules. The normal pulse width is several milliseconds long, but Q-switching can be used to produce pulses of tens of nanoseconds to microseconds and correspondingly higher peak powers.

The ruby laser is a three-level laser and as such requires a high threshold for lasing action to occur, since over 50% of the Cr^{3+} must be raised to the excited state to achieve population inversion. Several joules of energy per cubic centimeter are required to establish a threshold-level population inversion. A great deal of the energy from the Xe flash lamps used to excite the laser goes into heating the laser rod, consequently the duty cycle is limited. This problem is aggravated by the fact that ruby has a very low thermal conductivity. Cooling with refrigeration is used, but continuous operation is not practical.

The output of ruby lasers may be randomly or linearly polarized, depending on how the crystal is cut—that is, whether the c axis is parallel or perpendicular to the laser axis. The c axis parallel to the laser axis produces random polarization. When the c axis is perpendicular to the laser axis, linear polarization occurs.

The pulsed output of a ruby laser consists of random spiking due to heavily multimode operation. Q-switching is used to produce tens of megawatts peak power in 10- to 20-ns pulses. Ruby lasers are easily mode locked because of the multimode operation and gigawatts of peak power in 10-ps pulses can be achieved.

The ruby laser was the first to be used for piercing holes in diamonds for wire-pulling dies. It can be used for piercing small holes in many materials and for spot welding. It is also used in applications where short, Q-switched pulses of light in the visible are required. One such application is in pulsed holography. By using high-energy submicrosecond pulses from a ruby laser, holograms of moving objects can be made, thereby eliminating the need for careful prevention of even the slightest motion during exposure as required with low-power CW lasers.

8–6 NEODYMIUM-YAG LASERS

The neodymium (knee-oh-dim-e-umm) Nd-YAG laser uses triply ionized Nd as the lasant and the crystal YAG (yttrium-aluminum-garnet) as the host. YAG is a complicated oxide with the chemical composition $Y_3Al_5O_{12}$. The amount of metal ions replaced by Nd^{3+} is 1 to 2%. The output wavelength is 1.06 μm, which is in the near ir.

Nd-YAG lasers are capable of average power outputs up to 1000 W. YAG lasers are operated continuously up to a few hundred watts, but pulsed operation, at high repetition rates, is used at the higher power levels.

The YAG laser is a four-level system, which results in a lower threshold energy to achieve the necessary population inversion. Also, YAG has a relatively high thermal conductivity; so with cooling CW or high repetition pulsed operation is practical. CW YAG lasers are Q-switched with acousto-optical modulators to provide pulse rates of up to several thousand pulses per second, although average power begins to drop above 2000 pps.

Schematically the YAG laser head appears similar to that of ruby. Figure 8–5 is a photograph of a commercial Nd-YAG laser.

YAG lasers can be used for overlapping pulse seam welding, spot welding, hole piercing (including diamonds and other gemstones), and cutting. Cutting is accomplished by rapid overlapping hole piercing with a gas jet assist. Through frequency doubling and higher harmonic generation, YAG lasers are used for material property, photochemical, and interaction studies in the visible and ultraviolet ranges.

8–7 Nd-GLASS LASER

The Nd-Glass laser uses Nd^{3+} as the lasant, but the host material is glass. The output is still 1.06 μm, although the bandwidth is several times that of the Nd-YAG laser, which partially explains a much higher pulse energy output. The large bandwidth results in hundreds of axial modes operating simultaneously. Mode locking puts all these modes in step, timewise and spatially, thereby resulting in a large bandwidth, about 10^{12} Hz, and a correspondingly narrow pulse width of as low as 10^{-12} s.

High energy is achieved by using a physically larger active medium than in YAG lasers, which is practical with glass, for it can be made in excellent

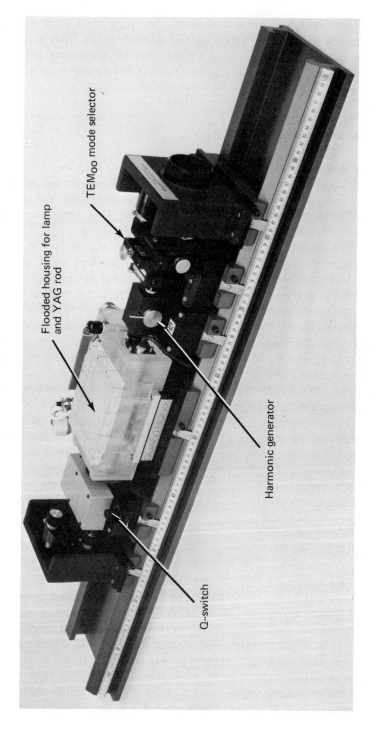

Flooded housing for lamp and YAG rod

TEM$_{00}$ mode selector

Harmonic generator

Q–switch

Figure 8–5 Nd-YAG laser. (Courtesy of Quantronix Corp.)

quality to almost any size. YAG crystals, on the other hand, are quite expensive and difficult to grow in large size. Unfortunately, glass has a low thermal conductivity; as a result, Nd-Glass lasers are limited to low-duty cycle operation of about one pulse per second. The energy per pulse can be 100 J or more, making the Nd-Glass laser suitable for many pulse-welding and hole-piercing applications.

One phenomenon occurring in solid lasers that can cause excessive beam divergence and self-focusing results from the fact that in some solids the refractive index increases with the irradiance of the beam. This creates a situation similar to that of a graded index optical fiber. The beam is actually guided by the self-induced refractive index charge. If the irradiance exceeds a threshold level, the beam is focused to a small spot, usually damaging or destroying the rod.

8-8 DIODE LASERS

Diode lasers, as might be guessed, are diodes that lase, not lasers that diode. So the phrase laser diode is incorrect (if anyone cares). There are about 20 known compound semiconductor diodes that lase with wavelengths ranging from 0.33 μm in the uv to nearly 40 μm in the ir. Electron beam excitation is used for all wavelengths shorter than 0.8 μm. In most of the others e-beam or injection current (current in a forward-based diode) can be used. For practical purposes, only the injection current type is discussed. The most common types of commercially available diode lasers are based on GaAs and ternary compounds, such as $Al_xGa_{1-x}As$. The $Al_xGa_{1-x}As$ compounds are discussed here as an example of diode lasers.

The first diode lasers operated in the early 1960s were simple homojunction GaAs devices, meaning that there was only one junction between the p- and n-type GaAs. Figure 8-6 is a schematic representation of a homojunction diode laser and the corresponding energy level diagram. The cavity is formed by the cleaved ends of the GaAs chip, which is about 0.5 mm wide and 0.1 mm high. Radiation at about 0.9 μm is emitted from the depletion region. Contact is made through gold, Au, deposited on the anode and the Au substrate that the chip is mounted on. Notice that the Fermi level is in the bands in the energy level diagram. This condition is indicative of degenerate (extremely heavy) doping approaching the solid solubility limit. Forward biasing of the diode results in a large injection current and a population inversion in the depletion region. The reason for the population inversion can be seen by inspection of Fig. 8-6. Only pulsed operation at liquid nitrogen temperature (77°K) is possible. Pulses with peak powers of a few watts are possible with repetition rates of several thousand pps.

Modern versions of the GaAs diode laser use single or double heterojunctions with stripe geometry and possibly large optical cavity (LOC) configurations to reduce the current density and risk of damage due to large radiation fields

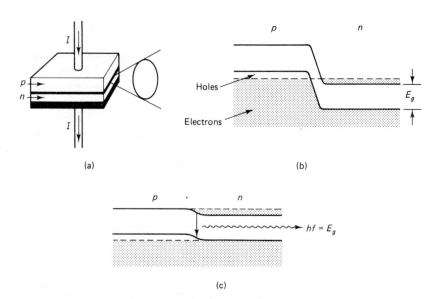

Figure 8–6 Homojunction diode laser (a) diode, (b) energy bands, unbiased, and (c) energy bands, forward biased.

in the chip. A heterojunction refers to a junction between GaAs and $Al_xGa_{1-x}As$. A double heterojunction device with stripe geometry is depicted in Fig. 8–7. The purpose of the stripe contact is to improve conduction of heat from the active region. The ternary compounds of $Al_xGa_{1-x}As$ have much poorer thermal conductivity than GaAs. Limiting the active region to a narrow stripe allows lateral heat conduction in GaAs to greatly reduce the temperature rise in the active region.

The purpose of the heterojunctions can best be explained in terms of the energy level diagram shown in Fig. 8–8. Note that the energy gaps are wider in $Al_xGa_{1-x}As$ than in GaAs. This factor has two effects. First, the resulting energy barriers reflect both electrons and holes back into the active GaAs region, thus reducing the current density required to produce stimulated emission. Secondly, there is a refractive index difference between $Al_xGa_{1-x}As$ and GaAs—3.4 for $x = 0.4$ and 3.6 for $x = 0$—which causes light traveling at slight angles to the cavity axis to be strongly reflected back into the active region so that less radiation need be generated, thereby lowering the required current density. Devices of this sort can be operated CW at room temperature with the aid of a thermoelectric cooler to prevent an excessive temperature rise. The current density required is 1000 A/cm². If the bottom heterojunction is replaced by an ordinary pn GaAs junction, the result is a single heterojunction device that can be operated in pulsed fashion at room temperature. Power levels are 10 to 20 W peak with repetition rates of up to 10^4 pps.

LOC devices are made with two additional layers of $Al_yGa_{1-y}As$ on either

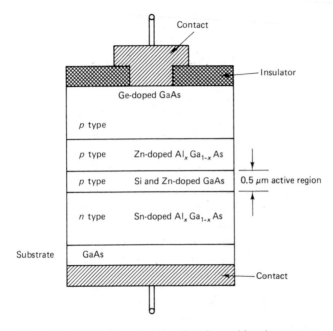

Figure 8-7 Double heterostructure diode laser with stripe geometry.

side of the *p*-GaAs with $y \ll x$ so that the optical cavity becomes much wider, tens of micrometers compared to perhaps 1 μm, thus reducing the danger of damage to the crystal from the radiation. This feature also greatly reduces the diffraction of the beam as it leaves the end of the crystal from an angle of as high as 30° to one of only about 2°. It should be mentioned that diode lasers achieve efficiencies of 4 to 5% for room temperature operation.

Variation of the amount of Al in these lasers permits adjustment of the emission wavelength from 0.84 to 0.95 μm.

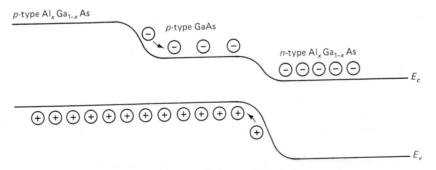

Figure 8-8 Energy level diagram for double heterostructure diode laser.

The largest application of diode lasers appears to be in the field of communications, particularly fiber optics communication. Because of the great speed with which diode lasers can be switched on and off and their narrow bandwidth, they are excellent candidates for high data rate fiber optic communications. Diode lasers will also undoubtedly find applications in pattern recognition, bar code reading, consumer products such as video disks, computer memory, and some ir illumination applications where fairly high-power monochromatic radiation is needed.

8–9 CO_2 LASERS

The carbon dioxide (CO_2) laser is a molecular laser in which molecular vibrations rather than electronic transitions provide the mechanism for lasing action. Other lasers in this category are the carbon monoxide and hydrogen fluoride lasers. The carbon monoxide laser has achieved high efficiency, around 40%, but the corrosiveness and toxicity of carbon monoxide have prevented scaling of carbon monoxide lasers to the power levels achieved in CO_2 lasers. Hydrogen fluoride lasers have been primarily of interest in gas-dynamic and chemical laser work aimed at extremely high-power levels, megawatts, for periods of time on the order of a second.

Carbon dioxide lasers have achieved success in a wide variety of industrial-materials-processing applications at power levels from a few watts to 15 kW with reasonably high efficiency of 10% for commercial models.[3] The types of materials that the CO_2 laser is applied to include paper, wood, glass, plastic, ceramics, and many metals. Industrial processes include heat treating, welding, cutting, hole piercing, scribing, and marking.

Because of its wide industrial use, the CO_2 laser is discussed in some detail. Some of the CO_2 laser designs that are commercially available are qualitatively described. Some specific CO_2 laser applications, along with material-processing applications with other types of lasers, are presented in Chapter 12.

General description.   Here we consider the interactions that take place in the CO_2 laser. Pure CO_2 can be made to lase but only weakly. Early in the development of CO_2 lasers, Patel discovered that a mixture of CO_2, N_2, and He greatly increased the power output. A typical ratio of $CO_2 : N_2 : He$ partial pressures is $0.8 : 4.5 : 10$. This will vary somewhat from system to system. The use of 10% oxygen mixed with the N_2 improves electrode and optics life without noticeably reducing power output.

The CO_2 molecule is a linear triatomic molecule and, as such, has only three distinct normal modes of vibration. These modes are shown schematically

---

[3] This is overall efficiency defined as the ratio of the useful power output to the *total* power input, not just discharge power.

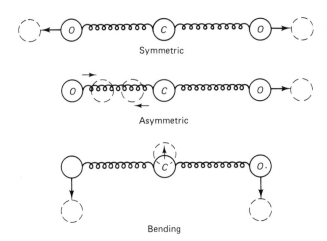

Symmetric

Asymmetric

Bending

Figure 8–9   Vibrational modes of the CO₂ molecule.

in Fig. 8–9. They are the asymmetrical stretching mode at $f_a = 7.0 \times 10^{13}$ Hz, the symmetrical stretching mode at $f_s = 4.2 \times 10^{13}$ Hz, and the bending mode at $f_b = 2.0 \times 10^{13}$ Hz. The bending mode is degenerate in that it can occur in either of two perpendicular planes.

Molecular vibrations, like electronic states, are quantized, meaning that energy can be added to or removed from these vibration states only in discrete quanta of energy. The amount by which the energy can change in any given vibration mode is given by $\Delta E = \pm hf$, where f is the frequency of the particular vibration mode. This leads to an energy level scheme for molecular vibrations similar to electronic energy level diagrams. A simplified version of the CO₂ energy level diagram is presented in Fig. 8–10, along with the first excited state of N₂. N₂ is a diatomic molecule; therefore it has only one normal mode of vibration. The abbreviated spectroscopic notation in Fig. 8–10 for CO₂ (l, m, n) refers to the number of quanta in the various vibration modes, l quanta in the symmetrical mode, m in the bending mode, and n in the asymmetrical mode. Only those levels actively involved in the excitation and lasing process are shown in Fig. 8–10.

Note that N₂ has a near coincidence between its first vibration state and that of the (001) state of CO₂. This excited state of N₂ is extremely long lived, nearly ensuring that it will give up its energy, received by means of collisions with electrons, to an unexcited CO₂ molecule in a resonant or elastic energy transfer. Direct excitation of CO₂ molecules by collision with electrons is also important. Lasing can occur between the (001) and (100) states at 10.6 μm or the (001) and (020) states at 9.6 μm. The probability for the latter transition is only about one-twentieth that of the 10.6 μm transition and so it does not occur normally unless the 10.6 μm transition is suppressed. From an industrial

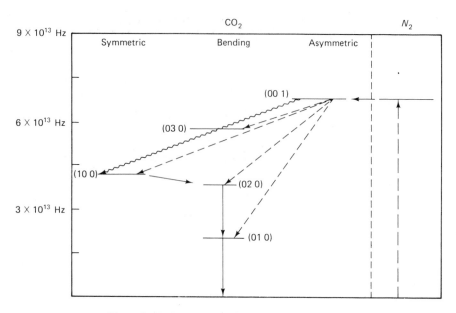

Figure 8–10   Energy level diagram for N_2 and N_2.

applications standpoint there is no advantage to operating at 9.6 μm, which would result in greatly reduced power output.

The vibrational states of the CO_2 molecule are split into many closely spaced energy states due to the quantized rotational states of the molecule. Quantum mechanical selection rules require that the rotational state either increase or decrease by only one quantum upon a transition from one vibrational state to another. That is, if J is the rotation quantum number, $\Delta J = \pm 1$. This situation leads to a large number of possible lines with two distinct branches, one for $\Delta J = 1$ centered at about 10.2 μm and one for $\Delta J = -1$ centered at 10.6 μm. Because of the slightly lower energy change for $\Delta J = -1$, this branch is favored, leading to a nominal output wavelength of 10.6 μm. Usually only one rotational transition operates because the one with highest gain is the most populated. Extremely rapid transitions between rotational levels occur due to thermal effects. The bandwidth of CO_2 lasers is about 60 MHz as a result of collisions and the Doppler effect.

The natural lifetime of the (001) state of CO_2 is of the order of seconds, but collision effects in the gas reduce this lifetime to the order of milliseconds, which is suitable for laser action. The collision lifetime of the lower states are on the order of microseconds, but thermal effects repopulate the (010) state, creating a bottleneck in the return of CO_2 molecules to the ground state. Collisions between CO_2 molecules in the (100) state and unexcited state can cause transfer of energy to the (020) state with subsequent collisions with unexcited

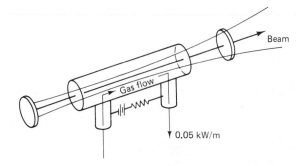

Axial discharge with slow axial gas flow

Figure 8–11 CO_2 laser with axial discharge and slow axial flow.

CO_2 molecules, causing transfer to the ground state, by collision with He atoms (which act as an internal heat sink), with the walls of a discharge tube, with a heat exchanger, or they are simply expelled from the active region. All CO_2 lasers of more than a few watts CW power output use some form of flowing gas technique to ensure a fresh supply of gas in the active region at all times. Many designs producing 1 kW or less simply exhaust the gas into the atmosphere. The total flow rate is around 0.1 SCMH.[4] The gas can be recirculated through a heat exchanger and is generally for higher-power level machines. A small amount of gas is still continuously or periodically removed and made up by fresh gas to keep the gas mixture free of decomposition products, particularly carbon monoxide, which absorbs strongly at 10.6 μm and will quench lasing action. Such byproducts are also harmful to the optics and electrodes.

CO₂ laser types.   Four basic types of CO_2 lasers are produced today for operation at 500 W CW or more. The differences in these types are significant enough that each is briefly described. Certain special laser designs, such as the TEA (transversely excited atmospheric pressure) laser, the waveguide laser, and other sealed tube types, will not be covered.

1. Axial discharge with slow axial flow.   This type of laser is schematically represented in Fig. 8–11. The power capability of this type is 50 to 70 W/m, although efforts are being made to increase this amount through better heat transfer from the gas to the cooling liquid that surrounds the laser tube in a separate tube or jacket. Power outputs up to 1.2 kW are achieved by placing tubes in series optically, with electrical discharge and gas flow in parallel. That is, the electrical discharge and gas flow are independent for each tube that is 2 or 3 m in length. The tubes are

---

[4] SCMH stands for standard cubic meters per hour.

Figure 8–12 One kilowatt axial flow laser. The laser head is in the large cabinet in back of the work station. (Courtesy of Saginaw Steering Gear, Division of GMC.)

usually mounted on a granite slab or in an Invar rod frame for mechanical and thermal stability. Figure 8–12 is a photograph of a 1-kW laser of this type.

Some of these lasers can be electronically pulsed to achieve much higher peak power without decreasing the average power significantly. This factor is advantageous in a number of material-removal operations. The tube diameter of such lasers is usually quite small, resulting in the TEM_{00} mode being strongly favored. Somewhat larger bore designs are available if low-order multimode operation is desired, such as for heat treating.

2. *Axial discharge with fast axial flow.* The chief difference between this design and the previous one is that the gas mixture is blown through the tubes at high speed and recycled through a heat exchanger. Each CO_2 molecule is in the tube only long enough to lase and then passes through the heat exchanger. The power output of this type of laser is 600 W/m. The tubes may be of fairly large bore; consequently, either unstable or higher-order mode outputs are common. Gaussian operation is reported for several kilowatts. Such lasers are available from 500-W to 6-kW CW power output with the potential for much higher power levels.

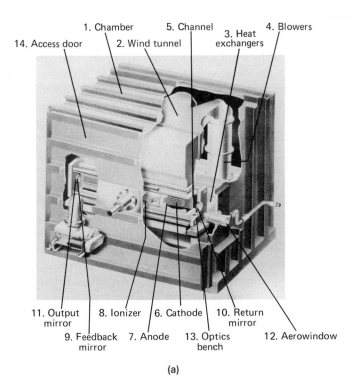

1. Chamber 5. Channel 4. Blowers
14. Access door 2. Wind tunnel 3. Heat exchangers

11. Output mirror 8. Ionizer 6. Cathode 10. Return mirror
9. Feedback mirror 7. Anode 13. Optics bench 12. Aerowindow

(a)

(b)

Figure 8–13 (a) CO_2 electron preionization laser with fast transverse flow. (b) Photograph of 15-kW CO_2 laser with electron beam preionization. (Courtesy of Avco Everett Research Laboratory, Inc.)

3. *Transverse electron beam preionization with fast transverse gas flow.*   It turns out that the electron energy required to sustain a discharge in a CO_2 laser is higher than optimum for production of a population inversion. One effective way to avoid this problem is to ionize the gas via a broad electron beam and then use a lower excitation voltage, which is not high enough to sustain the discharge but which provides optimum electron energy for excitation of the N_2 and CO_2 molecules. Figure 8–13 contains a schematic and a photograph of such a commercially available laser.

The power output of this type of laser reaches 10 kW/m and the commercial models provide up to 15 kW CW power. Because of the very high power level, these lasers are only operated with an unstable cavity. In the model shown an annular mirror placed around the back mirror, which is actually the output mirror, deflects the ring output past the side of the other mirror through an aerowindow. The aerowindow is a chamber with a vacuum pump attached to it. Air and laser gas entering the chamber from opposite ends are pumped out, thereby preventing air from entering the laser cavity.

The electron beam is produced in a high vacuum by the thermal emission of electrons from a large planar filament and acceleration through a high voltage. The primary electrons strike a thin metal foil separating the high vacuum of the electron gun and the much higher pressure in the laser cavity. Secondary electrons ejected from the foil produce the ionization in the laser gas that enables the low voltage discharge to be maintained. Such lasers have produced over 90 kW.

4. *Transverse discharge, transverse fast flow.*   This type of laser uses a fast transverse gas flow, about 60 m/s, with recirculation, like the previous design, but with a conventional electrical discharge transverse to the flow and the beam. The discharge takes place between a hollow water-cooled cathode or pins and a water-cooled segmented anode with each segment of the anode individually ballasted by a resistor. Relatively low voltage sustains the discharge at quite high current because of the short distance between the cathode and the anode. A power output of about 600 W/m is achieved. The beam is folded back and forth through the discharge region five to seven times, resulting in an actual power output of about 2.5 kW for a laser with approximately 1.2 m separating the cavity mirrors. Figure 8–14 shows a schematic diagram and a photograph of one such laser.

The outputs of these lasers can vary from nearly Gaussian to very high order or may be unstable, depending on the optical configuration of the cavity. Lasers of this type are available with power outputs from 1 to 9 kW and much higher power levels are attainable.

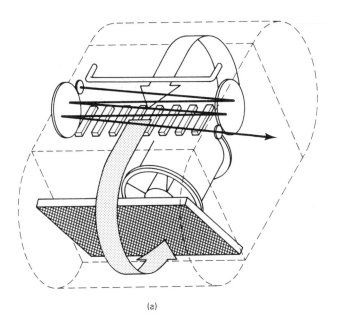

(a)

(b)

Figure 8–14 2.5-kW transverse discharge, transverse fast flow CO_2 laser. (a) schematic, (b) photograph. (Courtesy of Spectra-Physics, Industrial Laser Division.)

PROBLEMS

8–1. Discuss neutral gas, ion, and metal vapor lasers with particular emphasis on their differences.

8–2. Discuss dye lasers with particular emphasis on the pumping schemes and the reason for pulsed operation.

8–3. Discuss and compare ruby, Nd-YAG, and Nd-Glass lasers.

8–4. Discuss the diode laser with comparisons between homojunction and heterojunction devices and the methods used to reduce injection current density.

8–5. Discuss CO_2 laser operation and describe and compare three types of CO_2 laser designs.

8–6. Construct a table of pertinent data, such as wavelength, power or pulse energy range, bandwidth, and type of applications, for six of the lasers discussed in this chapter.

BIBLIOGRAPHY

8–1. Verdeyen, J. T., *Laser Electronics.* Englewood Cliffs, NJ: Prentice-Hall, Inc., 1981.

8–2. Svelto, O., *Principles of Lasers.* New York: Plenum Press, 1982.

8–3. Haus, Hermann A., *Waves and Fields in Optoelectronics.* Englewood Cliffs, NJ: Prentice-Hall, Inc., 1984.

8–4. Charschan, S.S. (ed.), *Lasers in Industry.* New York: Van Nostrand Reinhold Co., 1972.

8–5. Ready, J. F., *Industrial Applications of Lasers.* New York: Academic Press, 1978.

8–6. Silvast, W. T., "Metal-Vapor Lasers," *Scientific American,* February 1973.

8–7. Moss, T. S., G. J. Burrell, and B. Ellis, *Semiconductor Opto-Electronics.* New York: John Wiley & Sons, Inc., 1973.

8–8. Duley, W. W., *CO_2 Lasers: Effects and Applications.* New York: Academic Press, 1976.

8–9. Demaria, A. J., "Review of CW High-Power CO_2 Lasers," *IEEE Proceedings,* vol. 61, No. 6, June 1973.

chapter 9

Low-Power Laser Applications I: Alignment, Gauging, and Inspection

Industrial systems that use low-power lasers for alignment, gauging, or inspection generally consist of four basic units. These units are a laser light source, an optical system to direct and structure the light, an optical system to collect or image the light after it has interacted with the object or medium being interrogated, and a detection system. The detection system may be a human observer or an optoelectronic device interfaced with an electronic display unit or computer. This chapter describes the basic principles of low-power laser systems used in industry. Some comments on using a computer to interpret optical information are made, but basically this topic is beyond the scope of this book.

The applications of low-power laser systems discussed here do not depend on the coherent nature of laser light. Most systems are simple in nature and can be designed to operate in an industrial plant atmosphere. Low-power laser systems that depend on the coherent nature of laser light are discussed in Chapter 10. These systems tend to be more complex and often require the protection of a laboratory to operate in a reliable fashion.

9–1 SCANNING TECHNIQUES

Devices used to scan laser beams include mirrors and prisms. These devices are generally oscillated sinusoidally or rotated with a constant angular velocity. A mirror scanner is illustrated in Fig. 9–1. In many laser beam scanner applications the beam must remain parallel to a fixed axis as it scans. This can be accomplished by using a converging lens or mirror. A scanner system with a

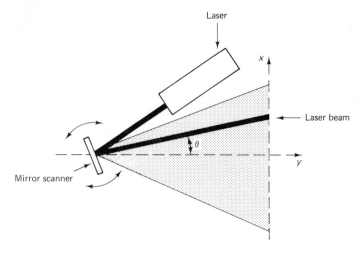

Figure 9–1 Mirror scanner.

converging lens is illustrated in Fig. 9–2. The scanner is used to scan the laser beam across a diameter of the converging lens. If the mirror is placed at the primary focal point of the lens, the beam will emerge from the lens parallel to the lens axis. Two element lens are often used to reduce spherical aberration to a minimum. If the lens is designed to meet the Abbe sin condition, the beam will emerge with a displacement

$$y = F \sin \theta \qquad (9\text{--}1)$$

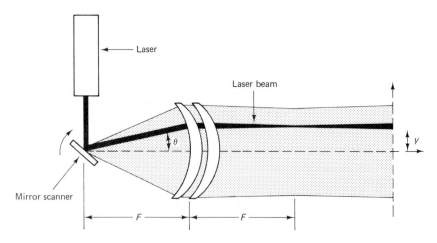

Figure 9–2 Beam scanner system with converging lens and mirror scanner at primary focal point of lens.

where F is the lens focal length and θ is the angular displacement of the beam entering the lens. θ is the scan angle. The scan angle θ, in terms of the angular speed ω of the mirror, is

$$\theta = 2\omega t \qquad (9\text{-}2)$$

Note that the angular speed of the beam is two times the angular speed of the mirror. The transverse speed v of the beam emerging from the lens is

$$v = \frac{dy}{dt} = 2\omega F \cos 2\omega t \qquad (9\text{-}3)$$

The transverse speed of the beam emerging from the lens is called the scan speed. For small angular displacements, Eq. (9-1) becomes

$$y = F\theta \qquad (9\text{-}4)$$

and Eq. (9-3) reduces to

$$v = 2\omega F \qquad (9\text{-}5)$$

Thus by using small angular displacements and rotating the scanner mirror with a constant angular speed, a constant scan speed can be obtained.

On emerging from the converging lens, the laser beam is not only parallel to the lens axis but also converges to a waist near the secondary focal plane of the lens. In many applications a converging beam is undesirable and a small-diameter, collimated beam is preferred. It can be attained by using a lens or mirror system or both, with a common focal point located at the reflecting surface of the scanner. Figure 9-3 is a sketch of a system that collimates the beam by using a large-diameter lens to focus the laser beam on a mirror scanner, where it is then reflected back through the same lens. In this case, the beam

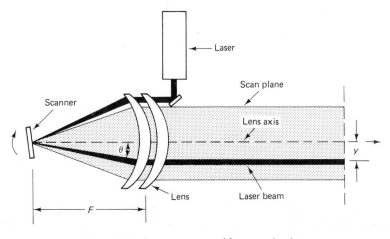

Figure 9-3 Scanner system with converging lens.

emerges from the system collimated, the same diameter as the beam directly from the laser, and the beam scans a plane parallel to the lens axis.

Concave and parabolic mirrors can also be used in scanner systems to produce a beam that remains parallel to a fixed axis. Parabolic mirrors eliminate spherical aberration whereas concave mirrors do not. Figure 9–4 is a sketch of a scanner system that uses a section of a parabolic mirror. The scanner is located at the focal point of the parabolic mirror. The light reflected from the parabolic mirror has a displacement of

$$y = 2F \tan\left(\frac{\theta}{2}\right) \qquad (9–6)$$

where F is the focal length of the parabolic mirror (see Problem 9–4). The scan velocity is

$$v = 2\omega F \sec^2(\omega t) \qquad (9–7)$$

As in the case of the previous system, Eq. (9–7) reduces to Eq. (9–5) for small angular displacements. In this system a lens with a focal point located at the scanner is used to collimate the laser beam.

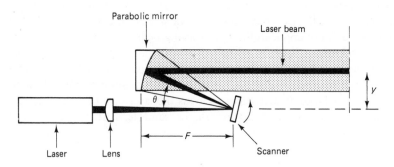

Figure 9–4 Scanner system with a parabolic mirror.

Shadow projection technique.   Objects should be backlighted whenever possible for inspection and gauging. Figure 9–5 is a simplified diagram of a system that uses a scanning laser beam to backlight an object so that its shadow can be used for gauging or inspection. This system uses an off-axis parabolic mirror to eliminate spherical aberration. A converging lens is used to focus the laser beam on a multisided mirror scanner placed at the focal point of the parabolic mirror. The beam is collimated after reflection from the parabolic mirror. By using an off-axis parabolic mirror, the rotating mirror can be placed so that it will not interfere with the reflected beam. The detector consists of a photodiode placed at the focal point of a second off-axis parabolic mirror. As the laser beam scans across an object placed between the scanner

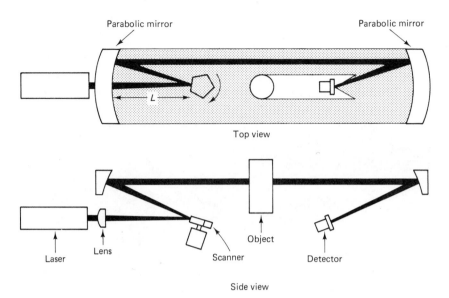

Figure 9-5   Simplified diagram of laser system designed for gauging or inspection.

and detector, the beam is blocked by the object for a time interval Δt. The shadow width Δy for small angular displacements, is

$$\Delta y = 2\omega L \, \Delta t \tag{9-8}$$

where L is the horizontal projection of the laser beam length from the scanner to the parabolic mirror. The output signal from the photodiode can be processed to determine Δt. This result can be electronically multiplied by $2\omega L$ and the product displayed on a digital readout. Accuracy of this system can be increased by focusing the beam to a narrow waist at the object position. Doing so reduces the time interval required for the cross section of the laser beam to sweep past an object edge. Figure 9-6 is a photograph of a commercial system being used to measure the diameter of a rod.

By using a computer in a laser scanner system, we are not limited to making measurements that require that the scan velocity be constant. Greater accuracy and measurements that require larger angular displacements can be obtained by integrating the scan velocity over the time interval that the laser beam is blocked by the object. In the previous example a computer can be used to evaluate the integral

$$\Delta y = 2\omega L \int_{t_1}^{t_2} \sec^2 \left(\omega t - \frac{\theta_0}{2} \right) dt \tag{9-9}$$

The time interval t_1 to t_2 is the time period that the laser beam is blocked by the object and θ_0 is the scan angle of the beam at time zero. The scan angle is the horizontal projection of the angle between the laser beam incident on

Figure 9–6 Laser scanner and detector being used to measure the diameter of a round bar. (Courtesy of Zygo.)

and reflected from the parabolic mirror. The computer can also be used to run statistical analyses and store data.

Surface inspection. Surface inspection with low-power lasers is done either by evaluating the specular or diffuse light reflected from the surface being interrogated. The laser beam used is almost always scanned so that the surfaces involved can be inspected in the shortest period of time. Evaluation of specular reflected laser light to detect surface defects has been highly successful for machined parts where the machined surfaces are specular and surface defects are filled with black residue from the machining process. Evaluation of diffuse laser light to determine differences in surface finish has been successful in some cases. One example is the determination of whether a machined surface has been honed.[1] Figure 9–7 illustrates a scanner system that is designed to detect surface defects and determine the presence of honed surfaces. If the machined surface is highly specular in nature, most of the laser light energy is reflected to the lens-photodiode detector positioned to receive the specular reflection. Very little light energy is diffusely reflected to a second detector positioned off to one side of the first detector. When the laser beam scanning a machined surface encounters a flaw, most of the energy is absorbed by the residue embedded in

---

[1] Honing refers to the machining of a surface with hone stones or emery cloth, which produces very fine periodic scratches.

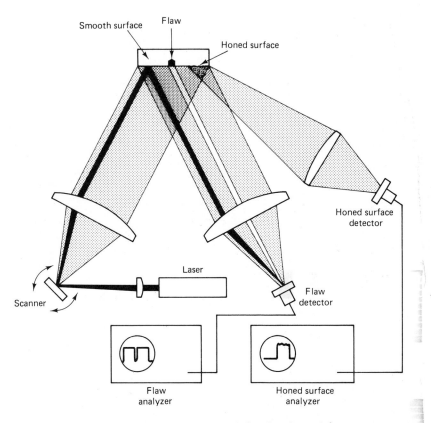

Figure 9-7   Laser scanner system designed to detect defects.

the flaw and specular reflected energy decreases. This drop in energy can be detected by an oscilloscope or electronics interfaced with the photodiode detector positioned to receive specular reflected light. To obtain a significant decrease in specular reflected energy, the spot size of the laser beam on the surface should be smaller than the smallest dimension of the flaw being interrogated. When scanned across a honed surface, the specular reflected light drops slightly and some of the light diffuses to the second detector. This increase of diffusely reflected light indicates the presence of a honed surface.

Scanner systems of this type can also be used to make measurements of hole diameters, object size, and edge locations and to determine the presence or absence of threads. Figure 9-8 is a diagram of a commercial scanner system designed to inspect for surface defects on sheets of material several meters wide. Here the detector is mounted above the sheet to observe diffusely reflected light. The detector consists of a pair of cylindrical lenses to focus the light on a glass or plastic tube. The back side of the tube is treated so it diffusely reflects the light that strikes it. The light entering the front side of the tube is diffusely reflected at the back surface and then transmitted, because of internal reflection,

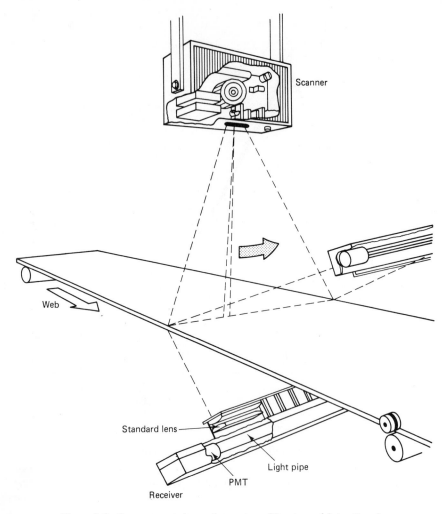

Figure 9–8   Laser scanner inspection system. (Courtesy of Intec Corp.)

down the length of the tube to a photomultiplier mounted on one end of the tube. A computer is used to process the output signals from the photomultiplier to indicate surface defects, such as pits, scratches, rust, and dark or rough spots. Transparent sheets can be inspected by mounting the detector so that it receives the light transmitted through the sheet.

If the reflectance of the object to be inspected can be controlled, or if it is highly uniform, the inspection system and evaluation of the information obtained is simple and straightforward. Experience has shown that industrial parts, even of the same type, can have a wide range of reflectance. Under these conditions signal processing is required to interpret properly the information obtained from the optical system. The systems discussed here are also very effective detectors of oil, dirt, ink marks, and other contaminants that do not represent surface defects. These surfaces must be clean in order to be properly inspected.

9–2 ALIGNMENT

Perhaps the most obvious application of a low-power laser is alignment. In this application the laser beam may simply be used as a pointer to align drainage tile, ditches, building foundations, or industrial equipment. By using a quadrant detector, objects can be aligned relative to a laser beam with high accuracy. A simplified diagram of an alignment system is illustrated in Fig. 9–9. If the laser beam is incident on the center of the detector, position indicators register midscale readings. When the beam is not incident on the center of the detector, the position indicators register a plus or minus reading, indicating to the user how to reposition the detector so that the beam is again centered.

If alignment is required over long distances, a beam expander can be used to reduce beam divergence. By expanding the laser beam, which reduces the irradiance, there is less potential hazard to the human eye. It is extremely important, even when working with low-power lasers, that current laser safety standards be followed and that lasers be used safely.

Laser beams can also be scanned to establish reference planes. This technique is particularly useful in the construction industry. Laser scanner systems are routinely used to install walls and ceilings. Figure 9–10 is a diagram of such a system. A beam expander is first used to increase the diameter of the

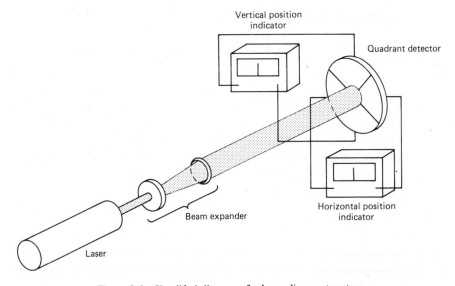

Figure 9–9 Simplified diagram of a laser alignment system.

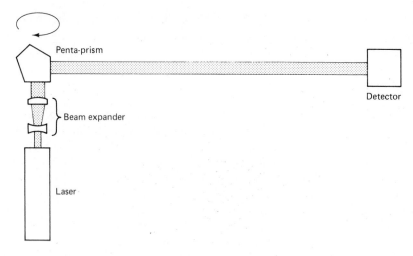

Figure 9–10   Diagram of a laser scanner system used for alignment.

laser beam. A penta-prism can be used to turn the beam at a right angle. A penta-prism turns the beam through a right angle even though the prism orientation is not precisely controlled. Rotating the prism about the axis of the incident laser beam enables the output beam from the prism to sweep out a flat plane. Detectors can be used to align objects relative to this plane. If high-precision alignment is not required, the spot of light produced when the laser beam strikes diffusive surfaces can be used directly for alignment.

9–3 TRIANGULATION TECHNIQUE

Optical triangulation provides a noncontact method of determining the displacement of a diffuse surface. Figure 9–11 is a simplified diagram of a laser-based system that is successfully used in many industrial applications. A low-power HeNe or diode laser projects a spot of light on a diffusive surface. A portion of the light is scattered from the surface and is imaged by a converging lens on a linear diode array or linear position detector. If the diffusive surface has a component of displacement parallel to the light incident on it from the laser, the spot of light on the surface will have a component of displacement parallel and perpendicular to the axis of the detector lens. The component of displacement perpendicular to the axis causes a corresponding displacement of the image on a detector. The displacement of the image on the detector can be used to determine the displacement of the diffusive surface.

Many triangulation systems are built with the detector perpendicular to the axis of the detector lens, as shown in Fig. 9–11. The displacement Δd of the image on the detector in terms of the displacement of the diffusive surface Δz, parallel to the incident laser beam, is approximately

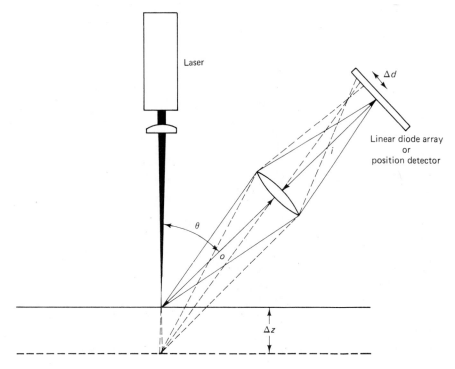

Figure 9-11   Diagram of laser-based optical triangulation system.

$$\Delta d = \Delta z \, m \sin \theta \qquad (9\text{-}10)$$

where $m = i/o$ is the magnification and θ is the angle between a line normal to the surface and the light scattered to the imaging lens. Equation (9-10) assumes that Δz is small and that θ remains constant as the surface is displaced. Thus we can use the approximate Eq. (9-10) to design an optical triangulation gauge; we cannot, however, assume that this equation is exact. After the gauge is built, it must be calibrated by displacing a diffusive surface by known distances and determining the actual image displacement on the detector. A computer then can be interfaced with the detector and programmed to determine the displacement of the diffusive surface from the displacement of the image on the detector.

The triangulation system illustrated is not designed to keep the image in focus as it is displaced on the detector. To keep the image in focus, the detector must be orientated at an angle to the lens axis as shown in Fig. 9-12.

Using Eq. (1-21), the image distance i is found to be

$$i = \frac{oF}{o - F} \qquad (9\text{-}11)$$

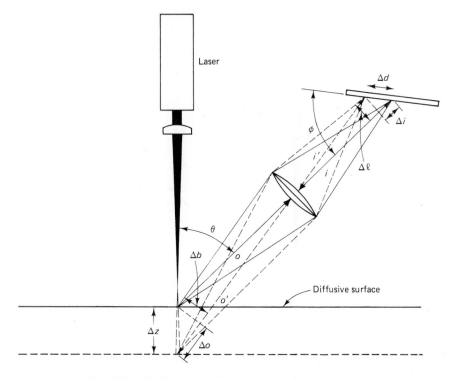

Figure 9–12 Diagram of optical triangulation system with detector at an angle relative to the lens axis.

where o is the object distance and F is the focal length of the lens. If the diffusive surface is displaced a distance Δz, in the direction of the incident laser beam, the new object distance o' is

$$o' = o + \Delta z \cos \theta \tag{9–12}$$

and the new image distance i' is

$$i' = \frac{oF + F\Delta z \cos \theta}{o - F + \Delta z \cos \theta} \tag{9–13}$$

The change in image distance is

$$\Delta i = i' - i = \frac{-F^2 \, \Delta z \cos \theta}{(o - F)^2 + (o - F) \, \Delta z \cos \theta} \tag{9–14}$$

For small displacements of the diffusive surface,

$$\Delta i = -\frac{F^2 \, \Delta z \cos \theta}{(o - F)^2} \tag{9–15}$$

To determine the angle of orientation ϕ of the detector, we will assume that

$$\tan \phi = \frac{\Delta l}{\Delta i} \tag{9-16}$$

where Δl is given by Eq. (9–10). That is,

$$\tan \phi = \frac{(\Delta z\, m \sin \theta)(o - F)^2}{F^2\, \Delta z \cos \theta} \tag{9-17}$$

where

$$m = \frac{F}{o - F} \tag{9-18}$$

Thus

$$\tan \phi = \frac{1}{m} \tan \theta \tag{9-19}$$

and

$$\Delta d = \frac{\Delta l}{\sin \theta} \tag{9-20}$$

or

$$\Delta d = m \left(\frac{\sin \theta}{\sin \phi}\right) \Delta z \tag{9-21}$$

Again, we cannot assume that Eq. (9–19) and Eq. (9–21) are exact equations. They are approximations that can be used to design an optical triangulation gauge. Then once built, the gauge must be calibrated to give maximum accuracy. Figure 9–13 is a photograph of an optical triangulation detector with a signal processor.

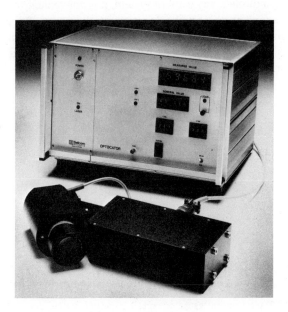

Figure 9–13   Laser-based optical triangulation system. (Courtesy of Selcom.)

9-4 DIODE-ARRAY CAMERA SYSTEM

Many optical inspection devices found in industry have a camera lens to focus the image of an object on a one- or two-dimensional diode array. White-light illumination is usually preferred in these applications. Laser illumination is often used when the light must be highly structured or directional or both. Unless otherwise stated, we will assume that a laser is being used as a light source. As in other applications, backlighting should be used whenever possible and reflected light used when necessary.

A typical diode-array camera system using laser light to backlight an object is illustrated in Fig. 9–14. A beam spreader collimates and increases the diameter of the laser beam. This laser light backlights an object that is to be inspected or gauged. A camera lens images the sharp edges of the object on a diode array. The output of the diode array can be displayed on an oscilloscope as illustrated in Fig. 9–14. Those diodes that are not illuminated by the image produce little or no signal and those diodes that are illuminated produce positive pulses as the diodes are scanned in sequence. By counting the number of diodes illuminated, or those not illuminated, or both, along with the magnification for the lens system, the dimensions of the object can be determined. If one-dimensional measurements are sufficient, then a linear diode array can be used. Two-dimensional measurements can be made by using a two-dimensional diode array. If a system of this type is used for gauging, then the lens must be designed so that a linear relationship exists between the object and image.

When using a diode-array camera system, it is convenient to predict mathematically the magnification and the location of the image for a given object distance. The camera lens must be treated as a thick lens. We have found the Newtonian form of the lens equations particularly useful for this purpose. The Newtonian form of the lens equation is

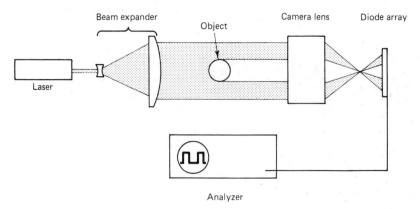

Figure 9–14 Typical optical gauge or optical inspection system designed by using a diode array camera.

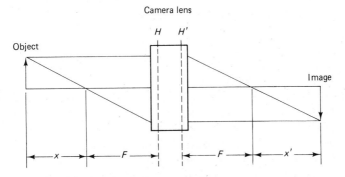

Figure 9-15   Notation for Newtonian form of lens equation.

$$xx' = F^2 \tag{9-22}$$

where x and x' are measured relative to the primary and secondary focal planes as shown in Fig. 9-15. The focal planes are easily found by passing a collimated laser beam through the camera lens in both directions and finding where the

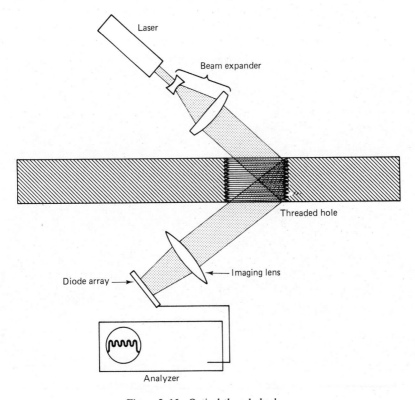

Figure 9-16   Optical thread checker.

beam focuses. The focal length of the lens can be obtained from the specifications on the lens. Magnification using the Newtonian form is

$$m = \frac{y'}{y} = -\frac{x'}{F} = -\frac{F}{x} \qquad (9\text{--}23)$$

The resolution with which a measurement of the image on the diode array can be made is plus or minus the distance between two adjacent photodiodes. Thus the theoretical resolution with which a measurement of the object can be determined is plus or minus the distance between adjacent photodiodes divided by the magnification.

Industrial parts can be inspected by frontlighting the parts if sufficient contrast is present so that the image can be interpreted by the diode array. Figure 9–16 shows a simplified diagram of an optical inspection system designed to detect the presence or absence of threads. If the threads are present, the output of the linear diode array is a sinusoidal waveform as illustrated. If threads are not present, the output for each diode tends to be the same and will produce a dc signal. This system could also be used to determine the pitch of the threads and to detect stripped threads. Rotating the object about the axis of the hole allows a complete inspection of the threaded hole to be made.

A photograph of a commercially available two-dimensional, diode-array camera system is shown in Fig. 9–17. Note that this system does not use a

Figure 9–17   Computer vision system. (Courtesy of Machine Intelligence Corporation.)

laser for illumination. Laser illumination is undesirable because of diffraction affects and speckle due to the coherent nature of laser light. When light of high irradiance must be directed from long distances into small, confined areas, however, a laser is often required.

PROBLEMS

9–1. Using $y = F\theta$, $y = F \sin \theta$, $y = 2\,F \tan (\theta/2)$ and assuming that $F = 20$ cm,
  (a) Plot beam displacement versus angular displacement for $0 \le \theta \le \pi/6$ on a single graph.
  (b) Find the beam displacement variation from $y = F\theta$ for $y = F \sin \theta$ and $y = 2F \tan (\theta/2)$ at $\theta = 3°$, $6°$, $12°$, $20°$, and $30°$.

9–2. Using $v = 2\omega F$, $v = 2\omega F \cos \theta$, and $v = 2\omega F \sec^2 (\theta/2)$ and assuming that $F = 20$ cm and $\omega = 2000\pi$ rad/s,
  (a) Plot beam velocity versus angular displacement for $0 \le \theta \le \pi/6$ on a single graph.
  (b) Find the beam velocity variation from $v = 2\omega F$ for $v = 2\omega F \cos \theta$ and $v = 2\omega F \sec^2 (\theta/2)$ at $\theta = 3°$, $6°$, $12°$, $20°$, and $30°$.

9–3. The beam scan angle θ for an oscillating mirror scanner is given by

$$\theta = \theta_{\max} \sin 2\pi f t$$

where f is the oscillating frequency and $\theta_{\max}$ is the angular displacement amplitude of the mirror.
  (a) Derive the equations for the transverse beam displacement and speed as a function of time for a laser beam emerging from a converging lens after being reflected by an oscillating mirror placed at the focal point of the lens.
  (b) Plot the beam displacement and velocity for $0 \le t \le 120$ μs if $F = 20$ cm, $f = 2000$ Hz, and $\theta_{\max} = 12°$.

9–4. The equation for a parabola is

$$z = \frac{y^2}{4F}$$

where F is the focal length. Assume that a laser beam originates from the focal point of a parabolic mirror at an angle θ with respect to the optical axis. Show that the laser beam is reflected from the mirror so its displacement y from the optical axis is

$$y = 2F \tan\left(\frac{\theta}{2}\right)$$

9–5. Integrate Eq. (9–9) to show that the displacement Δy of a laser beam for a time interval t_1 to t_2 for a system configured as shown in Fig. 9–5 is

$$\Delta y = \frac{2L \sin \omega(t_2 - t_1)}{\cos[\omega t_2 - (\theta_0/2)] \cos [\omega t_1 - (\theta_0/2)]}$$

Show that this equation reduces to Eq. (9–8) for small angular displacements.

9–6. Using Eq. (9–21), show that

$$\Delta d = m^2 \, \Delta z$$

as θ approaches zero and that

$$\Delta d = m \, \Delta z$$

for $\theta = \pi/2$.

9–7. Using Eq. (9–21), derive

$$\Delta d = m^2 \left(\frac{\cos \theta}{\cos \phi} \right) \Delta z = \left[\frac{F^2 \cos \theta}{(o - F)^2 \cos \phi} \right] \Delta z$$

9–8. An optical triangulation gauge is configured as shown in Fig. 9–12. The object distance is 12 cm, the image distance is 15.3 cm, and the angle between the laser beam and the optical axis of the lens is 40°. The detector is a 256 linear array with photodiodes placed at 25-μm centers.
   (a) Determine the focal length to the lens.
   (b) Determine the magnification.
   (c) Determine the angle between the diode array and the optical axis.
   (d) Determine the maximum displacement Δz that can be measured.
   (e) Using the distance between two photodiodes as a least count, determine the minimum displacement Δz that can be measured.

9–9. An optical triangulation gauge is configured as shown in Fig. 9–12 except the laser beam incident on the diffusive surface makes an angle β relative to a normal to the surface. Derive the equations

$$\tan \phi = \frac{1}{m} \tan (\theta + \beta) \qquad \text{and} \qquad \Delta d = \left[\frac{m \sin (\theta + \beta)}{\sin \phi \cos \beta} \right] \Delta z$$

which can be used to determine the detector angle ϕ and image displacement Δd for this gauge.

9–10. A diode-array camera gauge is configured as shown in Fig. 9–14. A 1024 linear array with photodiodes placed on 25 μm centers is used as a detector. Using the Newtonian form of the lens equation, determine the following for $x = 62.5$ mm and $x' = 48.4$ mm.
   (a) The focal length of the lens.
   (b) The magnification.
   (c) The number of diodes that would not be illuminated if the gauge is used to measure the diameter of a quarter-inch rod.
   (d) Determine the theoretical resolution of the gauge.

BIBLIOGRAPHY

9–1. Jenkins, F. A., and H. E. White, *Fundamentals of Optics*. New York: McGraw-Hill Book Co., 1976.

9–2. Sears, F. W., M. W. Zemansky, and H. D. Young, *University Physics*. Reading, MA: Addison-Wesley, 1976.

chapter **10**

Low-Power Laser Applications II: Interferometry and Holography

The purpose of this chapter is to provide the reader with a basic understanding of the laser-based optical systems in industry that depend on the coherent nature of light. An overview of some standard mathematical descriptions of laser light is presented first and then these descriptions are used to discuss applications of interferometry and holography.

10-1 MATHEMATICAL DESCRIPTIONS OF LASER LIGHT

One solution to the wave equation, Eq. (1–1), is the equation of a plane wave propagating along the z axis

$$E = \hat{i}E_0 \exp\left[j(kz - \omega t + \phi_0)\right] \qquad (10\text{--}1)$$

where E is the electric field intensity, E_0 the electric field intensity amplitude, $k = 2\pi/\lambda$ the propagation constant, $\omega = 2\pi f$ the angular frequency, j the complex number $\sqrt{-1}$, and ϕ_0 the original phase angle.[1] The wavelength of light is λ and f is its frequency. A unit vector $\hat{i}$ can be used in the equation when the light is plane polarized. To find the irradiance I in cases where E is the vector sum of two plane-polarized laser beams, the equation

$$I = \frac{<E \cdot E^*>}{2Z} \qquad (10\text{--}2)$$

[1] Boldface type denotes vector quantities and boldface italic type denotes complex vector quantities.

can be used. E^* is the complex conjugate of the electric field intensity. The angular brackets $<>$ denote the time average

$$<f(t)> = \frac{1}{T} \int_0^T f(t) \, dt \qquad (10\text{-}3)$$

T is the response time of the equipment that measures the irradiance or the exposure time for the film that records an image. In applications where the two laser beams are coherent the time-dependent term in Eq. (10–1) is not required to describe the light. To separate the time-dependent term, Eq. (10–1) can be written

$$E = \hat{\imath} E_0 \exp(jkz) \exp(j\phi_0) \exp(-j\omega t) \qquad (10\text{-}4)$$

Now it is possible to express Eq. (10–4) as

$$E = U \exp(-j\omega t) \qquad (10\text{-}5)$$

where U is the complex amplitude vector and is

$$U = \hat{\imath} E_0 \exp(jkz) \exp(j\phi_0) \qquad (10\text{-}6)$$

This complex amplitude vector is then used to describe a polarized plane wave because the time-dependent term is not required. It can be shown that for coherent linearly-polarized laser beams Eq. (10–2) reduces to

$$I = \frac{U \cdot U^*}{2Z} \qquad (10\text{-}7)$$

when $T \gg 1/f$.

Many other solutions to the wave equation are possible. Figures 10–1 and 10–2 are a summary of some possible solutions in rectangular and radial coordinates. Each solution can be expressed as a product of an amplitude, obliquity factor, original phase factor, and propagation factor. The amplitude term is constant in plane wave descriptions and a unit vector is used if the wave is linearly polarized. The propagation vector **k** has a magnitude of $2\pi/\lambda$ and is in the direction of propagation of the plane wave. To describe a plane wave propagating in a direction other than $+z$ directions, cosines γ_x and γ_y are used in the rectangular coordinate description and a position vector **r** is used in the radial coordinate description. The amplitude varies inversely proportional to r in mathematical descriptions of spherical waves. An obliquity factor is often used to describe the fact that laser light radiating from a point usually does not do so uniformly in all directions. The term $\cos\psi$ is used in Figs. 10–1 and 10–2, but it should be realized that it is not the only possible obliquity factor.

These mathematical descriptions should not be considered complete. They are intended to familiarize the reader with some descriptions that describe laser beams mathematically. To derive these equations is beyond the scope of this book. (For more complete presentations of this material, refer to Gaskill 1978 and Goodman 1968.)

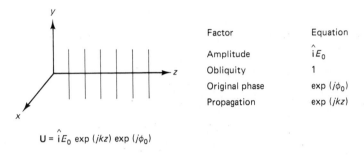

$$U = \hat{\mathbf{i}} E_0 \ \exp(jkz) \ \exp(j\phi_0)$$

(a) Polarized plane wave propagating in $+z$ direction.

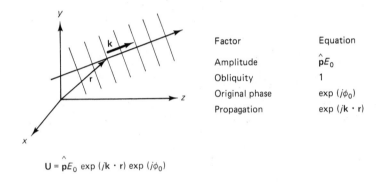

$$U = \hat{\mathbf{p}} E_0 \ \exp(j\mathbf{k} \cdot \mathbf{r}) \ \exp(j\phi_0)$$

(b) Polarized plane wave propagating in $+\mathbf{k}$ direction.

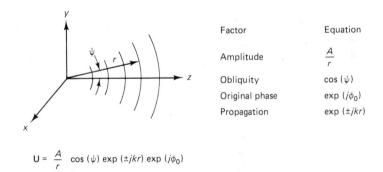

$$U = \frac{A}{r} \ \cos(\psi) \ \exp(\pm jkr) \ \exp(j\phi_0)$$

(c) Spherical wave diverging (positive exponent) or converging (negative exponent) with respect to origin.

Figure 10-1 Complex amplitude descriptions in rectangular coordinates.

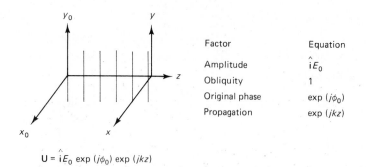

Factor	Equation
Amplitude	$\hat{i}E_0$
Obliquity	1
Original phase	$\exp(j\phi_0)$
Propagation	$\exp(jkz)$

$$U = \hat{i}E_0 \exp(j\phi_0) \exp(jkz)$$

(a) Polarized plane wave propagating in $+z$ direction.

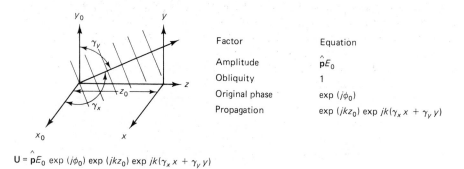

Factor	Equation
Amplitude	$\hat{p}E_0$
Obliquity	1
Original phase	$\exp(j\phi_0)$
Propagation	$\exp(jkz_0) \exp jk(\gamma_x x + \gamma_y y)$

$$U = \hat{p}E_0 \exp(j\phi_0) \exp(jkz_0) \exp jk(\gamma_x x + \gamma_y y)$$

(b) Polarized plane wave propagating in $+k$ direction.

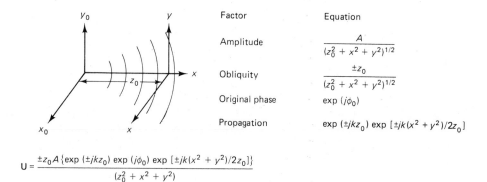

Factor	Equation
Amplitude	$\dfrac{A}{(z_0^2 + x^2 + y^2)^{1/2}}$
Obliquity	$\dfrac{\pm z_0}{(z_0^2 + x^2 + y^2)^{1/2}}$
Original phase	$\exp(j\phi_0)$
Propagation	$\exp(\pm jkz_0) \exp[\pm jk(x^2 + y^2)/2z_0]$

$$U = \frac{\pm z_0 A\{\exp(\pm jkz_0) \exp(j\phi_0) \exp[\pm jk(x^2 + y^2)/2z_0]\}}{(z_0^2 + x^2 + y^2)}$$

(c) Spherical wave diverging (positive exponent) or converging (negative exponent) with respect to the origin.

Figure 10–2   Complex amplitude descriptions in terms of radial coordinates.

10–2 LASER VELOCIMETRY

Consider two coherent plane waves with the same amplitude and polarization that intersect at an angle of ψ as shown in Fig. 10–3. In the region of intersection the interference fringes are bright and dark planes of light that are parallel to the plane bisecting the angle between the two beams. To derive the expression for the irradiance along the x axis, we will use the equation for a polarized plane wave propagating in the $+\mathbf{k}$ direction from Fig. 10–2. By assuming that $z_0 = \phi_0 = y = 0$, this equation can be expressed as

$$U = \mathbf{\hat{j}} E_0 \exp\left(jkx \cos \theta \right) \tag{10–8}$$

Thus the complex amplitude vectors of the two waves can be described by

$$U_1 = \mathbf{\hat{j}} E_0 \exp\left[jkx \sin \left(\frac{\psi}{2}\right) \right] \tag{10–9}$$

and

$$U_2 = \mathbf{\hat{j}} E_0 \exp\left[-jkx \sin \left(\frac{\psi}{2}\right) \right] \tag{10–10}$$

The resultant complex amplitude due to the interference of the two plane waves is

$$U = \mathbf{\hat{j}} E_0 \left\{ \exp\left[jkx \sin \left(\frac{\psi}{2}\right) \right] + \exp\left[-jkx \sin \left(\frac{\psi}{2}\right) \right] \right\} \tag{10–11}$$

Using Eq. (10–7), the irradiance I is given by

$$I = \frac{2E_0^2}{Z} \cos^2 \left[kx \sin \left(\frac{\psi}{2}\right) \right] \tag{10–12}$$

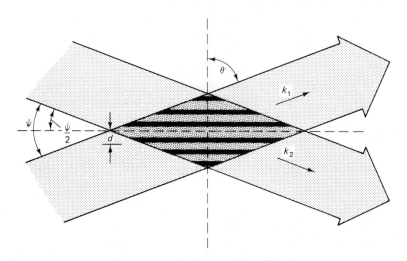

Figure 10–3 Intersecting coherent plane waves.

The distance d between two adjacent bright fringes can be found by letting

$$kd \sin \left(\frac{\psi}{2}\right) = \pi \tag{10–13}$$

Thus

$$d = \frac{\lambda}{2 \sin (\psi/2)} \tag{10–14}$$

The parallel fringes produced by two intersecting coherent laser beams provide a nonintrusive technique, called laser velocimetry, to determine the velocity of small particles moving in a transparent gas or liquid. One common technique is to seed a gas passing through a wind tunnel or engine with particles that will move with the same velocity as the gas. As the particles pass through the fringes, produced by the two intersecting beams, they scatter light proportional to the irradiance of the fringes. A sketch of a typical system is shown in Fig. 10–4. In this laser velocimetry system a laser beam is first split into two beams by a beam splitter. These beams pass through two holes in a mirror. Then they pass through a converging lens that focuses the beams to its focal point where they intersect, producing interference fringes. As a particle passes through the fringes, it scatters light back through the converging lens. A portion of the scattered light is reflected to a second converging lens that focuses this light on a photomultiplier. The signal from the photomultiplier is periodic and has a frequency

$$f = \frac{v_x}{d} \tag{10–15}$$

where v_x is the x component of the particle velocity that is perpendicular to the interference fringes. Other components of the velocity can be found by

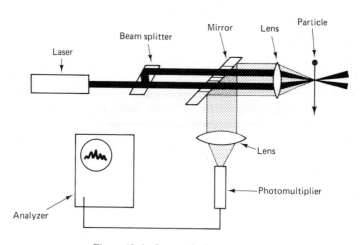

Figure 10–4   Laser velocimetry system.

forming fringes that are orthogonal to the first set. Signal separation is accomplished by using laser light of different colors or cross polarizing the light producing the fringe sets.

10–3 LASER INTERFEROMETRY

Industrial applications of laser interferometers include precise measurements of displacement, velocity, and surface variation between reflective reference and test surfaces. Long coherence length of the laser allows the reflective surfaces to be separated by large distances if necessary. In other techniques that do not use a laser, the test and reference surface must be in intimate contact. Linear displacement measurements can easily be made with an accuracy of one-quarter wavelength of the light being used. By using good measurement techniques and a computer system to compensate for changes in the index of refraction of air due to atmospheric effects, measurements can be made with even greater accuracy. As in the section on laser velocimetry, specific applications are discussed and mathematical descriptions are derived.

Michelson interferometer.   The Michelson interferometer illustrated in Fig. 1–22 can be used to make precise linear displacement measurements of one mirror relative to the other or to detect and/or measure surface variations between the two mirrors. We will first consider a Michelson interferometer system, as shown in Fig. 10–5, that uses a diverging lens to produce a diverging spherical wave. The beam splitter splits the diverging beam into two beams of equal irradiance. The two beams are reflected by the corresponding mirrors and equal portions are recombined by the beam splitter. These recombined

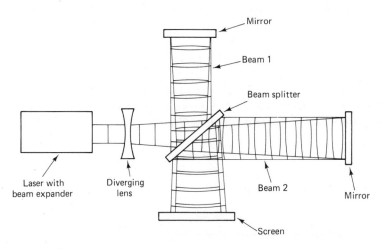

Figure 10–5 Michelson interferometer with diverging spherical waves.

waves interfere, producing fringes on a frosted glass screen that can be observed with the human eye.

The diverging wave can be described by the spherical diverging wave equation in rectangular coordinates from Fig. 10–2. Beam 1 can be described by the complex amplitude

$$U_1 = A \exp(jkz_1) \exp\left(\frac{jkr^2}{2z_1}\right) \tag{10–16}$$

and beam 2 by

$$U_2 = A \exp(jkz_2) \exp\left(\frac{jkr^2}{2z_2}\right) \tag{10–17}$$

We will assume that the amplitudes of the two beams at the screen are equal. Radii of the two beams are z_1 and z_2, respectively, and r is the radius of the pattern on the screen and is equal to $\sqrt{x^2 + y^2}$. Figure 10–6 is a better illustration

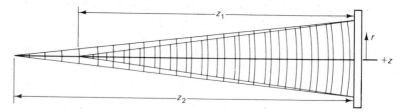

Figure 10–6 Superposition of two spherical waves with different radii propagating in the same direction.

of the two recombined beams. The two beams have different radii because of the different distances through which they travel. Resultant irradiance of the two interfering beams is

$$I = \frac{2A^2}{Z} \cos^2\left(\frac{k}{2}\right)\left[\Delta z + \frac{r^2}{2}\left(\frac{1}{z_1} - \frac{1}{z_2}\right)\right] \tag{10–18}$$

where Δz is the difference between the two radii z_1 and z_2. A photograph of the interference pattern produced is shown in Fig. 10–7. By observing the center of the pattern at $r = 0$, we can reduce Eq. (10–16) to

$$I = \frac{2A^2}{Z} \cos^2\left(\frac{k\,\Delta z}{2}\right) \tag{10–19}$$

Bright fringes are observed at $r = 0$ if

$$\frac{k\,\Delta z}{2} = m\pi \tag{10–20}$$

or

$$\Delta z = m\lambda \tag{10–21}$$

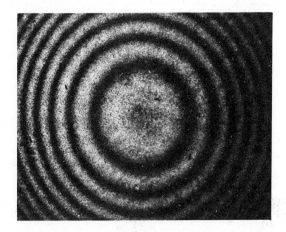

Figure 10–7 Photograph of the interference pattern due to two spherical waves with different radii propagating in the same direction.

where $m = 0,1,2,3,4, \ldots$. Dark fringes are observed at $r = 0$ if

$$\frac{k \, \Delta z}{2} = \left(m + \frac{1}{2} \right) \pi \tag{10–22}$$

or

$$\Delta z = \left(m + \frac{1}{2} \right) \lambda \tag{10–23}$$

Displacing one mirror of the interferometer a quarter wavelength in the direction of propagation of the light changes the optical path length of that light by a half wavelength. Thus at $r = 0$ the fringe would change from bright to dark or vice versa due to this displacement. That is, the least count for the displacement of a mirror in a Michelson interferometer is one-quarter wavelength of the light being used.

Variation between the two mirror surfaces can also be determined by using a Michelson interferometer. To check surface variation, the diverging lens is removed and a plane wave is used. If both mirrors are optically flat and one mirror is rotated through a small angle, straight, equally spaced fringes result as shown in Fig. 10–8. Surface variations are indicated by variation in the straightness and spacing of these lines. Figure 10–9 is a photograph showing the variation of these fringes due to a small bump on one mirror of a Michelson interferometer.

So that the observer can see the surface being tested as well as the interference fringes, an imaging lens is often added to a Michelson interferometer as shown in Fig. 10–10(a). Using this technique, the image of the surface tested and the interference fringes are superimposed on the screen. Interferometers of this type are called Twyman-Green laser interferometers and are often de-

Figure 10–8 Photograph of interference fringes due to the superposition of two plane waves propagating at an angle to each other.

signed to test curved surfaces as well as flat surfaces as illustrated in Fig. 10–10(b).

Fizeau interferometer. A Fizeau interferometer is illustrated in Fig. 10–11. In the next development we will use an optical flat for the reference surface and a convex mirror for the test surface. The reflection from the reference surface is a plane wave and the reflection from the test surface is a diverging spherical wave. The diverging spherical wave can be described by

$$U_1 = E_0 \exp\left(jkz_1\right) \exp\left(\frac{jkr^2}{z_1}\right) \tag{10–24}$$

Figure 10–9 Interference fringes due to a small bump on one of the mirrors on a Michelson interferometer.

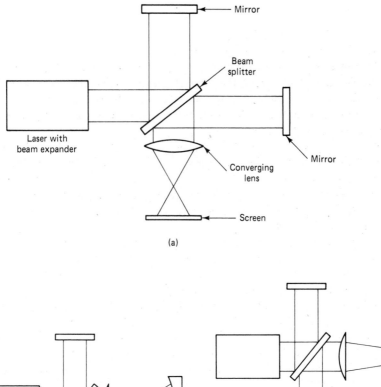

(a)

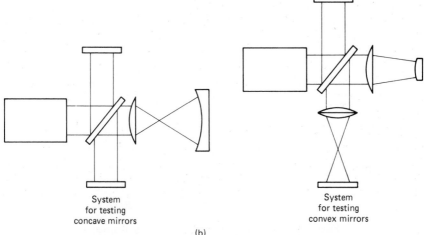

System
for testing
concave mirrors

System
for testing
convex mirrors

(b)

Figure 10–10   Schematic of a Twyman-Green laser interferometer.

The obliquity factor is assumed to be one, the amplitude E_0, the original phase angle zero, and z_1 is the radius of the spherical wave. The reflected plane wave from the reference surface can be described by

$$U_2 = E_0 \exp(j\phi_0) \exp(jkz_2) \qquad (10\text{–}25)$$

where z_2 is the distance from the focal point of the convex mirror where the plane wave has an original phase of ϕ_0. Solving for the irradiance by using Eq. (10–7), we have

$$I = \frac{E_0^2}{2Z} \cos^2 \left[\frac{1}{2} \left(\frac{kr^2}{2z_1} + k \, \Delta z - \phi_0 \right) \right] \qquad (10\text{--}26)$$

where r is the radius in the xy plane. We can find the radius of curvature of the test surface by imaging the convex surface on the screen with a magnification of one. First, we will adjust the interferometer so that the center spot in the resulting pattern is bright. We will then assume that $(k \, \Delta z - \phi_0)/2$ is $m_0\pi$, where m_0 is the order of the central bright fringe. Bright fringes will occur when

$$n = \frac{r^2}{\lambda z_1} \qquad (10\text{--}27)$$

where $n = m - m_0 = 0, 1, 2, 3, 4, \ldots$. The focal length of the spherical surface is z_1 and thus its radius of curvature R is equal to $2z_1$. Therefore

$$R = \frac{2r^2}{n\lambda} \qquad (10\text{--}28)$$

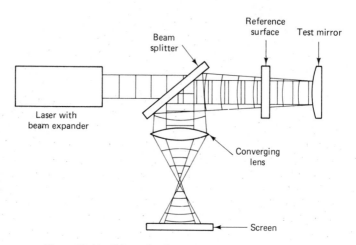

Figure 10–11 Schematic of a Fizeau laser interferometer.

As in the case of the Twyman-Green interferometer, the Fizeau interferometer can be designed to test flat surfaces as well as curved surfaces. Two-dimensional diode arrays interfaced with computers can interpret the fringes produced and quantify the results. A photograph of commercially available Fizeau interferometer is shown in Fig. 10–12.

Two-frequency interferometry. Another technique for making linear displacement measurements with an interferometer uses a laser that operates at two slightly different optical frequencies. A simplified schematic of such a system is illustrated in Fig. 10–13. When neon, the lasant, of a HeNe laser is exposed to a uniform magnetic field, Zeeman splitting takes place and the laser

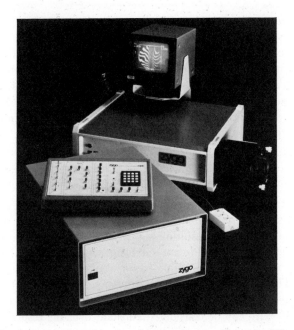

Figure 10–12   Fizeau laser interferometer system. (Courtesy of Zygo.)

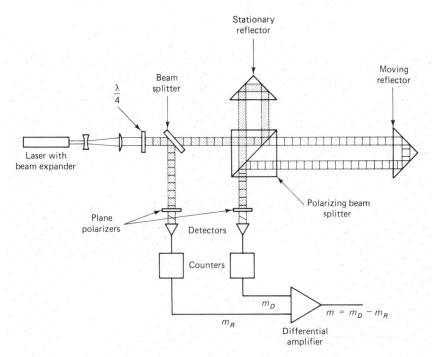

Figure 10–13   Simplified diagram of an interferometer system designed to measure linear displacement.

emits a laser beam of two frequencies, f_1 and f_2, with opposite circular polarizations. In the system illustrated, a quarter-wave plate is used to convert the oppositely circularly polarized signals into two orthogonal linearly-polarized signals. A beam splitter is used to obtain a small fraction of both linearly polarized signals, which are combined to obtain a reference signal. After being split from the primary beam, the orthogonal polarized reference signals are given the same polarization by a linear polarizer. By letting $z_1 = z_2 = 0$, the reference signals can be represented by

$$E_1 = \hat{n} E_0 \exp(-j\omega_1 t) \qquad (10\text{--}29)$$

and
$$E_2 = \hat{n} E_0 \exp(-j\omega_2 t) \qquad (10\text{--}30)$$

where ω_1 and ω_2 are the angular frequencies of the two signals. When the two reference signals are combined with the same polarization, the two signals beat. The frequency of the resulting signal is $(f_1 + f_2)/2$. The response time of light-measuring equipment is not sufficient to measure this high frequency. The resulting variation in amplitude has a frequency of $(f_2 - f_1)/2$ and can be detected. The resulting variation in irradiance that can be detected, due to the two signals, is found via Eq. (10–2). By using a high-speed detector, the time-dependent term can be considered constant over the response time T and the resulting irradiance is

$$I = \frac{2E_R^2}{Z} \cos^2 \frac{1}{2} (\omega_2 - \omega_1) t \qquad (10\text{--}31)$$

The frequency difference between the two signals is approximately 1.8 MHz. Irradiance has a maximum value when

$$\frac{1}{2} (\omega_2 - \omega_1) = m_R \pi \qquad (10\text{--}32)$$

where $m_R = 0, 1, 2, 3, 4, \ldots$, or when

$$m_R = (f_2 - f_1) t \qquad (10\text{--}33)$$

The reference signal is incident on a photodetector that produces an electrical signal that is proportional to the irradiance of the reference light. This electrical signal is amplified and used to drive a counter.

The primary beam passes through the reference beam splitter to a polarizing beam splitter. As shown in Fig. 10–13, the polarized light with frequency f_1 is transmitted through the polarizing beam splitter and the linearly-polarized light with frequency f_2 is reflected. The transmitted light is reflected back to the polarizing beam splitter by a cube corner reflector attached to an object that is experiencing linear displacement. As discussed in Section 1–9, light reflected from a moving reflector, which has a speed v much less than the speed of light c, is Doppler shifted by

$$\Delta f = \pm \frac{2fv}{c} \qquad (10\text{--}34)$$

The plus sign is used when the moving reflector moves toward the source of light and the negative sign when it moves away.

Light with frequency f_2, reflected by the polarizing beam splitter, is reflected back to the beam splitter by a second stationary cube corner reflector. This signal and Doppler-shifted signal are recombined at the polarizing beam splitter and then given the same polarization by a linear polarizer. After passing through the polarizer, and setting $z_3 = z_4 = 0$, the two signals can be represented by

$$E_3 = \hat{n} E_0 \exp\left[-j(\omega_1 \pm \Delta\omega)t\right] \tag{10–35}$$

and
$$E_4 = \hat{n} E_0 \exp\left(-j\omega_2 t\right) \tag{10–36}$$

The resulting irradiance I due to the two superimposed signals is

$$I = \frac{2E_0^2}{Z} \cos^2\left[\frac{1}{2}(\omega_2 - \omega_1 \pm \Delta\omega)t\right] \tag{10–37}$$

Now the irradiance can be expressed in terms of the optical frequencies f_1 and f_2.

$$I = \frac{2E_0^2}{Z} \cos^2\left[\pi(f_2 - f_1) \pm \frac{2f_1 v}{c} t\right] \tag{10–38}$$

Maximum irradiance occurs when

$$m_D = (f_2 - f_1)t \pm \frac{2f_1 v}{c} t \tag{10–39}$$

where $m_D = 0, 1, 2, 3, 4, \ldots$. As in the case with the reference light, the Doppler-shifted light is incident on a photodetector that produces an electrical signal proportional to the irradiance detected. This signal is amplified and a counter is used to make a count.

The counts from the reference and Doppler-shifted signal counters are subtracted, giving

$$m = m_D - m_R = \pm \frac{2f_1 v}{c} t \tag{10–40}$$

If counts are made over short time intervals, Eq. (10–40) becomes

$$\Delta m = \pm \frac{2f_1 v}{c} \Delta t \tag{10–41}$$

If we use $f_1/c = 1/\lambda$ and $v\,\Delta t = \Delta d$, Eq. (10–41) becomes

$$\Delta d = \pm \Delta m \frac{\lambda}{2} \tag{10–42}$$

The total displacement is obtained by making the summation

$$d = \sum_{i=1}^{N} \Delta m_i \frac{\lambda}{2} \tag{10–43}$$

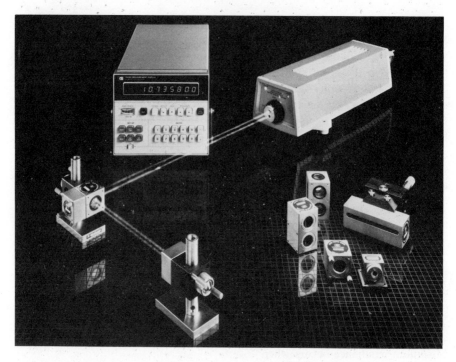

Figure 10–14 Laser measurement system. (Courtesy of Hewlett-Packard.)

A commercial laser measurement system that uses a two-frequency laser is shown in Fig. 10–14. Interferometers of this type can be used to measure linear displacement, velocity, flatness, pitch, yaw, straightness, squareness, parallelism, and perpendicularity.

10–4 HOLOGRAPHIC INTERFEROMETRY

Dennis Gabor published the first papers on holography in 1948. From 1948 through 1960 progress in the use and development of holography was slow because researchers working in this discipline did not have a light source available that could produce intense light with a long coherence length. Emmett Leith and Juris Upatnieks, working at the University of Michigan Institute of Science and Technology in the early 1960s, produced the first holograms using a laser. In doing so, Leith and Upatnieks transformed holography from an obscure concept to a viable scientific and engineering tool. In 1965 Powell and Stetson published the first paper on holographic interferometry. Dennis Gabor received the Nobel Prize in physics for his work in 1972.

For our purposes, a hologram can be thought of as an optical device—produced by using photographic techniques and laser light—that is capable of creating three-dimensional images. The word *hologram* stems from the Greek root *holos*, which means whole, complete, or entire, and *gram*, which means message. Thus a hologram is a complete record of an optical scene. In conventional photography the light reflected from a scene is focused—using a converging lens system—onto a photographic emulsion. Variation of the irradiance due to the image being focused on the emulsion is related only to the electric field intensity amplitude of the light. In holography a photographic emulsion, usually on a glass plate, is exposed to an interference pattern produced by two coherent laser beams. One beam, called the object beam, is reflected from an object or scene to the photographic emulsion. The other beam, called the reference beam, is reflected directly from the laser to the photographic emulsion by using mirrors. Using this technique, both the amplitude and the phase information about the electric field due to light reflected from the scene can be recorded. After photographic development, the resulting transparency is a hologram. By shining laser light—in some cases, white light—through the hologram, three-dimensional images can be produced.

Recording process.   Figure 10–15 is a sketch of an optical system that can be used to produce holograms. All the optical elements used in the production of optical holograms are mounted on an essentially vibration-free surface. Usually a heavy table with a steel or granite top, isolated from floor vibration by air tubes, provides this surface. A photograph of a typical hologra-

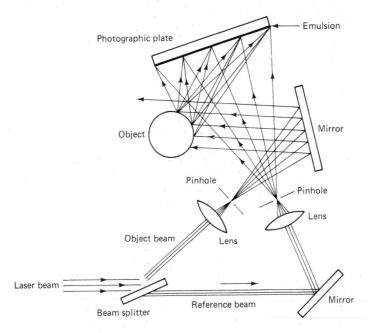

Figure 10–15 Basic optical system for producing holograms.

Figure 10–16   Typical holography system.

phy system is shown in Fig. 10–16. A continuous or pulsed laser operating in the TEM$_{00}$ mode at a wavelength within the visible portion of the electromagnetic spectrum is used for most industrial applications. Helium-neon, argon, and krypton are continuous-wave lasers commonly used for holography. Pulsed ruby lasers are used to produce holograms of transient phenomena or when vibration is a problem.

To produce a hologram, a laser beam is first split into two beams by a beam splitter as shown in Fig. 10–15. The reference beam is reflected by a mirror and then spread by a converging lens to illuminate uniformly a photographic emulsion on a glass plate. The object beam is spread by a converging lens and then reflected to the object by a mirror so that the object is uniformly illuminated as seen from the position of the photographic plate. Pinholes are used in the focal planes of the converging lenses as spatial filters to remove optical noise from the laser beams. The optical noise is primarily due to light scattered by dust and flaws on the optical element surfaces. Maximum coherence of the two beams at the photographic plate is obtained by making the object and reference beams of equal length.

It is desirable that objects used in holography have diffusive surfaces. Flat white paint is often used to give industrial parts uniform diffusive surfaces. When illuminated, each point on these surfaces will act as point sources, scattering light in all possible directions. Thus optical information about every point on the object facing the photographic plate can be recorded on an infinite number of points on the photographic emulsion.

Consider the complex amplitude of the laser light scattered from a point on the object to a point on the photographic emulsion to be U_O and the complex amplitude due to the reference beam at the same point on the emulsion to be U_R, as illustrated in Fig. 10–17. The resultant complex amplitude of the light at the point on the emulsion is the vector sum

$$U = U_R + U_O \tag{10–44}$$

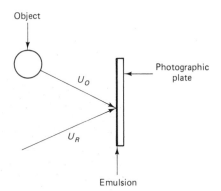

Figure 10-17   Recording configuration.

Using Eq. (10–7), the irradiance is found to be

$$I = \frac{1}{2Z} (U_R \cdot U_R^* + U_O \cdot U_O^* + U_R^* \cdot U_O + U_R \cdot U_O^*) \qquad (10\text{–}45)$$

This result can be rewritten

$$I = I_R + I_O + \frac{1}{2Z} (U_R^* \cdot U_O + U_R \cdot U_O^*) \qquad (10\text{–}46)$$

where the first two terms are the irradiances of the reference and object beams, respectively. The third term depends on the relative phase and polarization of the beams. If the object and reference beams are cross polarized

$$I = I_R + I_O \qquad (10\text{–}47)$$

all phase information is lost and a hologram cannot be produced. Ideally, then, both beams should be linearly polarized in the same direction. We will assume that such is the case for the remaining discussion on holographic interferometry.

Exposure $\mathscr{E}$ is the product of the irradiance I and the exposure time T.

$$\mathscr{E} = IT \qquad (10\text{–}48)$$

After development, a typical emulsion used in holography has a light amplitude transmittance versus exposure before development as shown in Fig. 10–18. To obtain a linear recording, the emulsion is "biased" to establish a quiescent point in the linear portion of the transmittance versus exposure curve. By making a linear recording, it can be assumed that the amplitude transmittance of the emulsion after development is a linear function of the irradiance incident on the emulsion during exposure. That is,

$$t = t_0 + \beta \left(I_0 + \frac{U_R^* U_O}{2Z} + \frac{U_R U_O}{2Z} \right) \qquad (10\text{–}49)$$

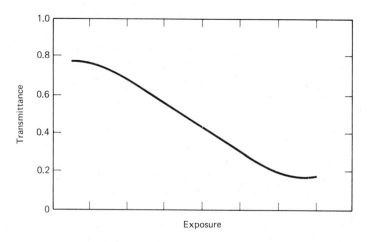

Figure 10–18 Light-amplitude transmittance versus exposure for typical holographic emulsion.

where t_0 is the "bias" transmittance established by the reference beam and β is the product of the slope of the transmittance versus exposure curve at the quiescent point and the exposure time [Goodman 1968].

Reconstruction. Holograms produced by using the optical system diagrammed in Fig. 10–15 can be used to form virtual and real images. To "play back" a virtual image, a reconstruction beam is used that is identical to the reference beam used in the production of the hologram. In most industrial applications a laser is used for reconstruction. When a reconstruction beam, which is identical to the reference beam, is used to illuminate the hologram, much of the light passes straight through the hologram. Some light is diffracted by the fringes recorded in the emulsion due to its exposure to the interference pattern produced by the object and reference beams.

An observer viewing the diffracted light, as illustrated in Fig. 10–19, will see a virtual image of the object in space behind the hologram. This image is three dimensional and in the same location as the object was relative to the emulsion. The image will appear to be the same as if the observer were looking at the original object, illuminated by the light used for reconstruction, through a window the size of the hologram.

Assume that a hologram, produced in the manner discussed, is illuminated by a reconstruction beam U_P. The light transmitted through the hologram is

$$U_P t = U_P t_0 + \beta \left(U_P I_0 + \frac{U_P U_R^* U_O}{2Z} + \frac{U_P U_R U_O^*}{2Z} \right) \qquad (10\text{–}50)$$

It is traditional to write Eq. (10–50) as

$$U_P t = U_1 + U_2 + U_3 + U_4 \qquad (10\text{–}51)$$

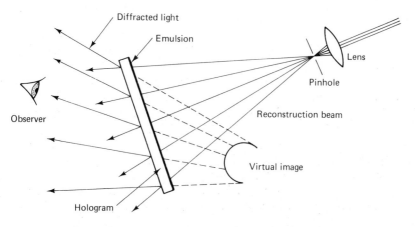

Figure 10–19   Holographic reconstruction of a virtual image.

where
$$U_1 = U_P t_0 \qquad (10\text{–}52)$$

$$U_2 = \beta U_P I_0 \qquad (10\text{–}53)$$

$$U_3 = \frac{\beta U_P U_R^* U_O}{2Z} \qquad (10\text{–}54)$$

and
$$U_4 = \frac{\beta U_P U_R U_O^*}{2Z} \qquad (10\text{–}55)$$

To reconstruct the virtual image,

$$U_P = b U_R \qquad (10\text{–}56)$$

That is, the reconstruction beam is identical to the reference beam except for a multiplicative constant b. In this case, it can be shown that the light transmitted through the hologram distributes itself as indicated in Fig. 10–20.
Note that

$$U_3 = (b\beta I_R) U_O \qquad (10\text{–}57)$$

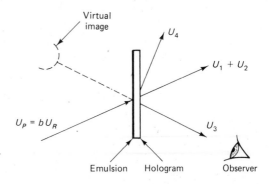

Figure 10–20   Virtual image reconstruction.

Thus the light represented by this term is an exact duplicate of the object beam except for the multiplicative constant $b\beta I_R$. The light represented by this term is identical to the light reflected from the object to the emulsion when the hologram was produced!

To reconstruct a real image using holography, it is common practice to collimate the reference beam. This being the case, the reconstruction system in Fig. 10–21 can be used to reconstruct a real image of the object. In this playback process, a collimated reconstruction beam is used to illuminate the hologram from the opposite direction that the reference beam had when the hologram was produced. This image appears in front of the hologram and is formed in the same location, relative to the emulsion, as the location of the object when the hologram was produced.

To reconstruct the real image,

$$U_P = bU_R^* \tag{10–58}$$

The reconstruction beam is the complex conjugate of the reference beam multiplied by the constant b. For a collimated reference beam, the complex conjugate of U_R represents a collimated beam traveling in the opposite direction of the original reference beam. (See Problem 10–8.) In this case, the light transmitted through the hologram distributes itself as indicated in Fig. 10–22. Using this playback scheme,

$$U_4 = (b\beta I_R)U_O^* \tag{10–59}$$

Light represented by this term is an exact duplicate of the object beam except for the multiplicative constant $b\beta I_R$ and the fact that the light is conjugated. This conjugated light produces a real image of the object.

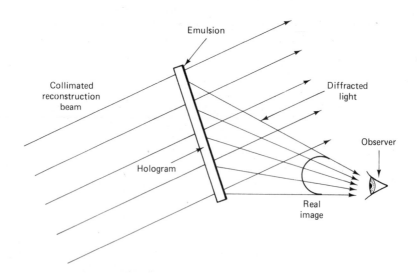

Figure 10–21 Holographic reconstruction of a real image.

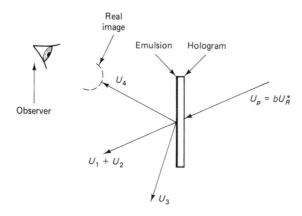

Figure 10-22   Real-image reconstruction.

Real-time holographic interferometry.   Figure 10–15 illustrates a holography system that is frequently used for all types of holographic interferometry. In real-time holographic interferometry the photographic plate is exposed, developed, and then placed back in the same position it occupied during exposure. By replacing the developed plate (hologram) and turning the laser back on, the object beam then illuminates the object, and the reference beam becomes the reconstruction beam for the hologram. The object is then viewed through the hologram. The observer will see the object and the virtual image reconstructed by the hologram. If all the elements in the holography system have maintained the same position, the virtual image and the object will be located in the same position. Now if the object experiences a small surface deformation, fringes will be observed due to the interference of the light reflected by the object and the light diffracted by the hologram that forms the virtual image. These fringes are similar to the lines on a topographical map that are superimposed on the object. These fringes can be used to determine surface displacement.

Surface displacement is most easily determined when a collimated object beam is used and the object beam, surface displacement, and light scattered to the hologram are all directed along an axis that is normal to the object surface as illustrated in Fig. 10–23. The object illustrated in this figure is a circular diaphragm mounted in a rigid frame. The diaphragm is flat when the photographic plate is exposed and then deformed so it forms a convex surface relative to the object beam after the hologram is returned to the holography system.

Assume that the original object beam is

$$U_O = a(x, y) \tag{10-60}$$

Assume that a hologram of the object has been produced and it has been returned to the holography system so that it can be reconstructed by the reference beam.

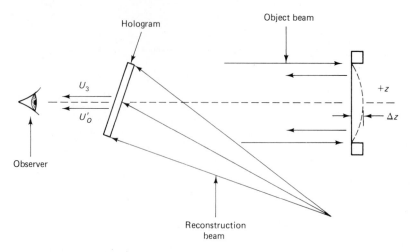

Figure 10–23 Simple holographic interferometry system with object beam normal to object surface.

When reconstructed, the light producing the virtual image of the object can be expressed as

$$U_3 = -a(x, y) \qquad (10–61)$$

where the multiplicative constant $b\beta I_R$ is assumed to be -1. The negative sign is required for most real-time holographic interferometry because the photographic plates usually used in this work have negative emulsions, making β negative. In this reconstruction process, the light reflected from the object and viewed through the hologram can be expressed as

$$U'_O = a(x, y) \exp [j \, \Delta\phi(x, y)] \qquad (10–62)$$

$\Delta\phi(x, y)$ is the phase shift due to the displacement Δz. Irradiance due to the interfering beams is

$$I(x, y) = \frac{2a^2(x, y)}{Z} \sin^2 \left[\frac{\Delta\phi(x, y)}{2} \right] \qquad (10–63)$$

Note that when $\Delta\phi(x, y)$ is zero, the image and object are in the same position and the irradiance is zero. Change in object beam path length is two times the surface displacement for the configuration being discussed. Hence

$$\Delta\phi(x, y) = \frac{2\pi}{\lambda} [2 \, \Delta z(x, y)] \qquad (10–64)$$

Dark fringes correspond to phase changes $\Delta\phi(x, y) = 2\pi m$, where $m = 0, 1, 2, 3, 4, \ldots$. If a section of the object surface experiences no displacement, dark fringes can be assigned numbers $m = 0, 1, 2, 3, \ldots$. where zero is

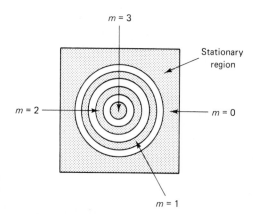

Figure 10–24   Real-time holographic image due to normal displacement of a circular diaphragm.

assigned to the stationary surface. By counting the number of dark fringes from the stationary surface, the displacement along the z axis can be calculated using

$$\Delta z(x, y) = m \frac{\lambda}{2} \qquad (10\text{–}65)$$

An illustration of the fringes observed using the real-time holography technique discussed is shown in Fig. 10–24. Recall that the outside perimeter of the diaphragm was held stationary and that the diaphragm cups away from the observer as it is displaced. Using Eq. (10–65), it is obvious that the center of the diaphragm has been displaced $3\lambda/2$.

Unfortunately, not all surface displacements can be determined via this simple technique. Complex object shapes require a vector analysis to determine surface displacement. Techniques for calculating more complex displacements are discussed in the next section. It should be realized, however, that the more generalized techniques discussed in the next section can also be applied to real-time holographic interferometry.

Example Problem

Find the displacement of the diaphragm center, illustrated in Fig. 10–24, if the blue line of an argon ion laser is used.

Solution. We have assumed that the displacement is in the $+z$ direction. Counting the dark fringes starting with zero at the stationary surface, we determine that m is 3. The blue line from an argon laser has a wavelength of 488 nm. Using Eq. (10–65), we have

$$\Delta z(0, 0) = \frac{(3)(488 \text{ nm})}{2}$$

Thus

$$\Delta z(0, 0) = 732 \text{ nm}$$

Double-exposure holographic interferometry.   The holographic system illustrated in Fig. 10–15 can also be used for double-exposure holographic interferometry. Here the emulsion is exposed with the object in its original state, the object is deformed, and then a second exposure is made on the same emulsion. The emulsion is developed after the second exposure. The resulting hologram can be reconstructed by using the reference beam or its duplicate. Two images will be reconstructed. One image is of the object before deformation and the second is of the object after deformation. As in the case of real-time holographic interferometry, fringes will be observed that can be used to determine the surface displacement that took place between the two exposures.

Fringes observed in double-exposure holographic interferometry are due to the interference of the light that produces the holographic images. Figure 10–25 can be used to determine the equations needed to determine surface displacement. Point O is the location of the converging lens focal point in the object beam. Point P is the original position of a point on the object surface. Its final position after surface deformation is P' and $\mathbf{d}$ is the surface displacement. Images reconstructed by the hologram are observed through point Q. Because surface displacement must be small, it can be assumed that the rays represented by $\mathbf{r}_1$ and $\mathbf{r}_1'$ are parallel. The same is true for $\mathbf{r}_2$ and $\mathbf{r}_2'$. $\mathbf{k}_1$ and $\mathbf{k}_2$ are propagation vectors for the light illuminating P and light scattered to Q, respectively. Phases of the rays observed in reconstruction are

$$\phi_1 = \mathbf{k}_1 \cdot \mathbf{r}_1 + \mathbf{k}_2 \cdot \mathbf{r}_2 \tag{10–66}$$

and

$$\phi_2 = \mathbf{k}_1 \cdot \mathbf{r}_1' + \mathbf{k}_2 \cdot \mathbf{r}_2' \tag{10–67}$$

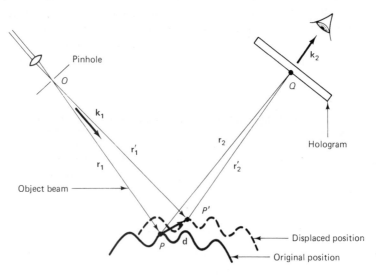

Figure 10–25   Ray diagram for evaluating surface displacement.

It is assumed that the original phase of the two rays at point O and the phase shift due to reflection from the object surface are both zero. The relative phase of the two rays at point Q is

$$\Delta\phi = \phi_2 - \phi_1 = \mathbf{k}_1 \cdot (\mathbf{r}_1' - \mathbf{r}_1) - \mathbf{k}_2 \cdot (\mathbf{r}_2 - \mathbf{r}_2') \qquad (10\text{–}68)$$

Vector differences $\mathbf{r}_1' - \mathbf{r}_1$ and $\mathbf{r}_2 - \mathbf{r}_2'$ are both equal to the surface displacement **d**. Thus

$$\Delta\phi = (\mathbf{k}_1 - \mathbf{k}_2) \cdot \mathbf{d} \qquad (10\text{–}69)$$

To demonstrate the use of Eq. (10–69), let us consider a circular diaphragm with the object beam originating from a point source and at an angle relative to the z axis as shown in Fig. 10–26. The vector difference

$$\Delta\mathbf{k} = \mathbf{k}_1 - \mathbf{k}_2 \qquad (10\text{–}70)$$

is parallel to the bisector of the angle between $\mathbf{k}_1$ and $-\mathbf{k}_2$ and has a magnitude of $2k \cos(\psi/2)$. Therefore

$$\Delta\phi = \Delta k\, d \cos\theta \qquad (10\text{–}71)$$

or
$$\Delta\phi = 2kd \cos\left(\frac{\psi}{2}\right) \cos\theta \qquad (10\text{–}72)$$

where θ is the angle between the vector difference $\Delta\mathbf{k}$ and the surface displacement **d** and ψ is the angle between $\mathbf{k}_1$ and $-\mathbf{k}_2$. Note that the sensitivity of the system is determined by both the angle θ and the vector difference magnitude Δk. For simplicity, θ and $\psi/2$ are equal in Fig. 10–26. It should be realized

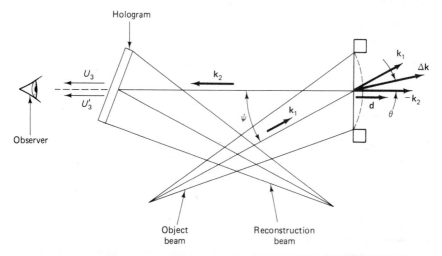

Figure 10–26 Simple holographic interferometer configuration with object beam illuminating object surface at an angle relative to surface normal.

that this may not be true for other configurations. Equation (10–72) reduces to Eq. (10–64) when the surface displacement and the object beam direction before and after reflection from the object surface are all parallel to the same axis. To determine **d** with a single observation, the angle θ must be known.

In completing the analysis of our circular diaphragm, assume that

$$U_3 = -a(x, y) \tag{10–73}$$

and

$$U_3' = -a(x, y) \exp [j \, \Delta\phi(x, y)] \tag{10–74}$$

where U_3 and U_3' are the complex amplitudes of the light reconstructing the images due to the first and second exposures, respectively. The resulting irradiance due to the interfering light is

$$I(x, y) = \frac{2a^2(x, y)}{Z} \cos^2 \left[\frac{\Delta\phi \, (x, y)}{2} \right] \tag{10–75}$$

Unlike the real-time case, the irradiance is maximum when the relative phase between the two beams $\Delta\phi(x, y)$ is $2\pi m$, where $m = 0, 1, 2, 3, 4, \ldots$. Figure 10–27 is a sketch of the resulting image that would be observed by using double-exposure holographic interferometry and the holography system illustrated in Fig. 10–26. The surface displacement can be calculated by counting bright fringes from the stationary portion of the object, using

$$d = \frac{m\lambda}{2 \cos (\psi/2) \cos \theta} \tag{10–76}$$

Recall that θ is the angle between the vector difference $\Delta\mathbf{k}$ and the surface displacement **d**. ψ is the angle between $\mathbf{k}_1$ and $-\mathbf{k}_2$ or the angle between the object light illuminating a displaced point on the object surface and the light scattered from that point to the observer.

Example Problem

Find the displacement of the diaphragm center illustrated in Fig. 10–27 if a HeNe laser is used in the holography system and $\psi/2 = \theta = 22°$.

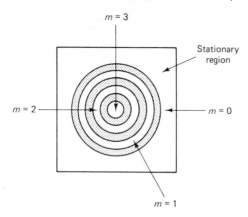

m = 3

Stationary region

m = 2 ——————— m = 0

m = 1

Figure 10–27 Double-exposure holographic image due to normal displacement of a circular diaphragm.

Solution. Using Fig. 10–27, m is 3 and λ for a HeNe laser is approximately 633 nm. Using Eq. (10–76), we have

$$d = \frac{3(633 \text{ nm})}{2 \cos^2 22°}$$

Thus

$$d = 1.10 \ \mu\text{m}$$

When the direction of the displacement is not known, then the measurement that can be easily determined is the component of the displacement parallel to the vector difference $\Delta\mathbf{k}$—assuming that a stationary portion of the object exists and can be observed. Obviously, three components of the displacement can be determined by making three observations from three different angles. This step can be accomplished by using three different holograms or by making three different observations through a single hologram. This situation indicates that the fringe location relative to an object changes as an observer views the displaced surface from different angles. Such, in fact, is the case. The fringes can exist in a plane other than the object surface. (A more extensive coverage of holographic interferometry is given by Vest 1979.)

Displacement measurement in many industrial applications is not as important as knowing which portion of an object surface experiences the greatest displacement between the two exposures of a hologram. Figure 10–28 is a photograph of a double-exposure hologram image that was produced with a holography

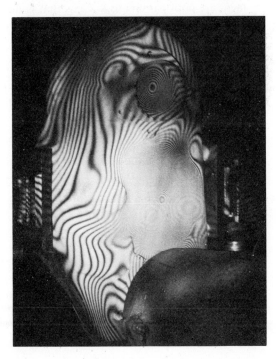

Figure 10–28 Photograph of reconstructed image from a double-exposure hologram produced using a double-pulsed ruby laser. (Courtesy of Detroit Diesel Allison Division, General Motors Corporation.)

system using a double-pulsed ruby laser and reconstructed with a HeNe laser. Here the outside case of a gear box was vibrating at one of its resonant frequencies at a particular operating condition. It is obvious that the surface area located on the right side of the gear box is vibrating with maximum amplitude. To reduce vibration and thus the noise produced by the outer surface of this gear box, for the operating condition investigated, this portion of the surface must be treated to reduce its motion.

Vibration analysis is one of the most common industrial applications of holography. Most other techniques of vibration analysis require that a device, such as an accelerometer, be mounted on a vibrating surface. Holography is a noncontact technique. When the object to be analyzed is small, holography offers distinct advantages over other techniques.

Time-average holographic interferometry. Many laboratories do not have a pulsed laser system that can be used for vibration analysis by double-exposure holographic interferometry. The alternative is time-average holographic interferometry, which can be accomplished using a low-power continuous wave laser.

To simplify our analysis, we will assume, as in the discussion of real-time holographic interferometry, that a plane wave is used to illuminate the object. Furthermore, we will assume that the propagation direction of the light before and after being scattered from the object surface and the surface displacement are both parallel to a z axis that is normal to the object surface as shown in Fig. 10–23. In this analysis, however, we will assume that the circular diaphragm is vibrating sinusoidally such that

$$z(x, y, t) = D(x, y) \cos \omega_D t \qquad (10\text{--}77)$$

where $D(x, y)$ is the amplitude of vibration at position (x, y) on the diaphragm and ω_D is the angular frequency of vibration. The corresponding phase change due to this vibration is

$$\Delta\phi(x, y, t) = 2kD(x, y) \cos \omega_D t \qquad (10\text{--}78)$$

The complex amplitude of the object beam at the hologram will be

$$U_O(x, y, t) = A(x, y) \exp [2jkD(x, y) \cos \omega_D t] \qquad (10\text{--}79)$$

We will assume that the period of vibration of the diaphragm is much smaller than the exposure time. Information recorded by the hologram due to the object will be the time average

$$U_O(x, y) = \frac{1}{T} \int_0^T A(x, y) \exp [2jkD(x, y) \cos \omega_D t] \, dt \qquad (10\text{--}80)$$

where T is the exposure time. This integration yields

$$U_O(x, y) = A(x, y)J_0[2kD(x, y)] \qquad (10\text{--}81)$$

where J_0 is a Bessel function of the first kind of order zero.

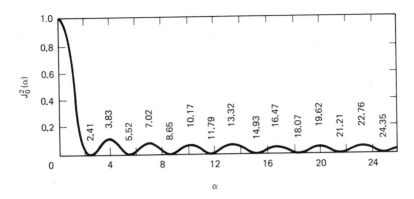

Figure 10–29 Plot of $J_0^2(\alpha)$ versus α.

When the hologram is developed and reconstructed with a reconstruction beam that is identical to the original reference beam, the complex amplitude of the light producing the virtual image is

$$U_3 = (b\beta I_R)A(x, y)J_0[2kD(x, y)] \qquad (10\text{–}82)$$

The irradiance of this light is

$$I(x, y) = \frac{(b\beta I_R)^2}{2Z} A^2(x, y)J_0^2[2kD(x, y)] \qquad (10\text{–}83)$$

Figure 10–29 is a plot of $J_0^2(\alpha)$ versus α. Note that the positions on the object that do not vibrate are easily determined because they are brighter than other regions where higher-order bright fringes exist. The virtual image of a sinusoidally vibrating circular diaphragm, vibrating at its fundamental resonant frequency, is illustrated in Fig. 10–30. The outside perimeter of the diaphragm is very bright, indicating that this region is stationary. The dark fringes corre-

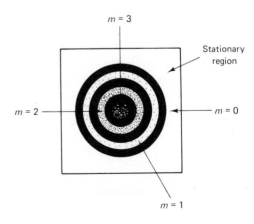

Figure 10–30 Virtual holographic image of a circular diaphragm vibrating sinusoidally at its fundamental resonant frequency.

spond to the zero values of the Bessel function. The bright fringes have a decreasing irradiance as the order of the fringes increase. That is, the center bright spot on the diaphragm illustrated in Fig. 10–30 has less irradiance than the circular bright fringe adjacent to it and each larger diameter fringe is brighter.

Example Problem

Determine the amplitude of vibration of the circular diaphragm shown in Fig. 10–30 if a HeNe laser operating at a wavelength of 633 nm is used to produce the time-average hologram. As in the preceding development, assume that the holographic system used is as illustrated in Fig. 10–23.

Solution. The central bright spot is a third-order bright fringe. From Fig. 10–29 we can determine that

$$2kD(0, 0) = 10.2$$

Thus

$$D(0, 0) = \frac{(633 \text{ nm})(10.2)}{4\pi}$$

and

$$D(0, 0) = 514 \text{ nm}$$

Example Problem

Figure 10–31 shows the ray diagram of a time-average holography system and virtual image produced with this system that is used to determine the mode and amplitude of vibration of a rod fixed at one end. The rod displacement and direction of the object beam light scattered to the photographic plate are both along the z axis, which is normal to the equilibrium position of the rod. The angle between the object beam light illuminating the rod and light scattered to the photographic plate is 30°. Determine the mode of vibration and vibration amplitude of the rod tip if a HeNe laser operating at 633 nm is used to produce the hologram.

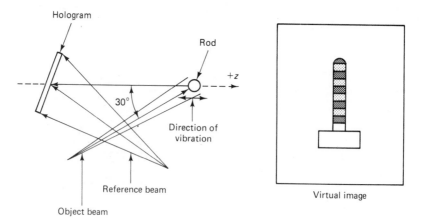

Figure 10–31 Ray diagram of a time-average holography system and virtual image produced with the system to evaluate a vibrating rod.

Solution. The bright rod base indicates that it is stationary. No other areas of equal brightness are present. Therefore the rod is vibrating at its fundamental resonant frequency. A dark fringe is located at the rod tip. Using Fig. 10–29, we determine that this dark fringe corresponds to $\alpha = 14.93$. Using Eq. (10–72) and substituting α for $\Delta\phi$ and D for d, we obtain

$$D = \frac{14.93\lambda}{4\pi \cos(\psi/2) \cos\theta}$$

Both $\psi/2$ and θ are equal to $15°$. Thus the amplitude of vibration of the rod tip is 806 nm.

PROBLEMS

10–1. Given the expression for the electric field intensity

$$E = \frac{10\ \mu V}{m} \cos\left(\frac{10^7}{m}z + \frac{3 \times 10^{15}}{s}t + 0.75\right)$$

find the
(a) Amplitude
(b) Propagation constant
(c) Wavelength
(d) Angular frequency
(e) Frequency
(f) Original phase angle
(g) Direction of propagation
(h) Irradiance in free space by using Eq. (1–7)

10–2. Given the expression for the complex amplitude of the electric field intensity due to a point source

$$U = \frac{100\ \mu V/m}{r} \cos(\psi) \exp\left[j\left(\frac{1.22 \times 10^7}{m}r\right)\right]$$

find the
(a) Amplitude
(b) Propagation constant
(c) Wavelength
(d) Frequency
(e) Original phase angle
(f) Irradiance at $\psi = 15°$ and $r = 1$ m

10–3. Determine the maximum and minimum distance between adjacent bright fringes due to two interfering laser beams that can be described as coherent plane waves.

10–4. Derive the expression for the irradiance I as a function of z due to two interfering plane waves

$$U_1 = \hat{\imath}E_0 \exp(jkz)$$

and $$U_2 = \hat{\imath}E_0 \exp(-jkz)$$

Plot irradiance I as a function of z and determine the distance between adjacent bright fringes. Repeat the exercise for the two plane waves

$$U_1 = \hat{\imath} E_0 \exp (jkz)$$

and
$$U_3 = \hat{\jmath} E_0 \exp (-jkz)$$

10–5. Show that the expression for the radius of the dark fringes shown in Fig. 10–7 is

$$r = \sqrt{\frac{2z_1z_2}{m_0 + \frac{1}{2}}} \sqrt{n}$$

where m_0 is the order of the central dark fringe and $n = m - m_0$, where m is the fringe order. Measure the radii of the dark fringes in Fig. 10–7. Then use log-log paper or a power curve-fitting computer program to show that the equation for the dark fringe radius has the form

$$r = an^b$$

Determine a and b.

10–6. Two laser beams with opposite circular polarizations can be described by

$$E_1 = A(\hat{\imath} + j\hat{\jmath}) \exp (-j\omega t)$$

and
$$E_1 = A(\hat{\imath} - j\hat{\jmath}) \exp (-j\omega t)$$

Describe the beam produced by the superposition $E_1 + E_2$ and the resulting irradiance. Repeat the exercise for two beams with opposite circular polarizations that have slightly different optical frequencies.

10–7. Using the count versus time data given in Fig. 10–32, determine the displacement of the moving reflector of a two-frequency interferometer. Plot the moving reflector

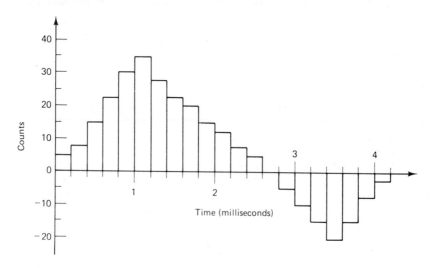

Figure 10–32   Graph for Problem 10–7.

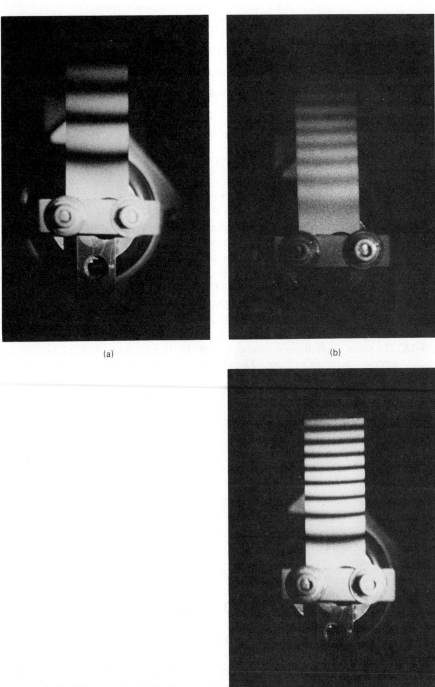

Figure 10–33 Photographs for Problems
10–13 and 10–14.

displacement versus time. Determine the velocity of the moving reflector for each time interval and plot a velocity versus time curve.

10–8. Given the expression for the complex amplitude

$$U = A \exp (jkz) \exp (j\phi)$$

find the direction of propagation of $E_1 = U \exp (-j\omega t)$ and $E_2 = U^* \exp (-j\omega t)$.

10–9. Given the expression for the complex amplitude

$$U = A \exp(jkz_0) \exp \left[\frac{jk(x^2 + y^2)}{2z_0} \right]$$

describe this wave and its complex conjugate.

10–10. Describe the results if a hologram is reconstructed with the object beam.

10–11. Find the displacement of the diaphragm center illustrated in Fig. 10–24 if the angle between the object beam and surface normal before and after it is scattered from the diaphragm is in the same plane and equal to 30°.

10–12. Using the holography system and virtual image illustrated in Fig. 10–31, determine the order and type of fringe located at the tip of the vibrating rod if a double-pulsed ruby laser was used to produce a double-exposure hologram with the two exposures taken with the rod in positions of maximum positive and negative displacement relative to equilibrium.

10–13. The series of photographs shown in Fig. 10–33 were taken during a holographic interferometry experiment. The object is a rectangular steel plate clamped at one end.
  (a) State which is a photograph of the image due to a real-time, double-exposure, and time-average hologram.
  (b) State the reasons for your conclusions.

10–14. The holography system geometry used in the experiment described in Problem 10–13 is identical to the one illustrated in Fig. 10–31, except $\psi = 22°$.
  (a) Determine the maximum displacement of the steel plate by using the double-exposure and real-time hologram photographs shown in Fig. 10–33.
  (b) Determine the maximum vibration amplitude of the steel plate by using the time-average hologram photograph.

REFERENCES

GASKILL, J. D., *Linear Systems, Fourier Transforms, and Optics*. New York: John Wiley & Sons, Inc., 1978.

GOODMAN, J. W., *Introduction to Fourier Optics*. New York: Academic Press, Inc., 1968.

VEST, C. M., *Holographic Interferometry*. New York: John Wiley & Sons, Inc., 1979.

BIBLIOGRAPHY

10–1. Caulfield, H. J. (ed.), *Handbook of Optical Holography*. New York: Academic Press, Inc., 1979.

10–2. Collier, R. J., C. B. Burckhardt, and L. H. Lin, *Optical Holography*. New York: Academic Press, Inc., 1971.

10–3. Fowles, G. R., *Introduction to Modern Optics*. New York: Holt, Rinehart & Winston, Inc., 1968.

10–4. Snyder, J. J., "Laser Wavelength Meters," *Laser Focus*, May 1982, pp. 55–61.

chapter 11

Interaction
of High-Power Laser Beams
with Materials

Several basic concepts of energy balance and heat transfer, as they apply to laser processing, are presented in this chapter. The developments given are intended to provide understanding and, in some cases, a reasonable approximation (at least order of magnitude) to reality. The types of applications to which the concepts developed here are applicable are heat treatment, welding, and material removal. Alloying, cladding, and glazing, less important industrially at this time, can also be treated.

The interaction of a laser beam with the workpiece depends on several laser beam parameters as well as material parameters. These parameters are discussed prior to a treatment of energy balance and heat transfer concepts.

11–1 MATERIAL AND LASER PARAMETERS

Material and laser parameters that affect laser processing are discussed in this section.

Material parameters

1. Reflectance R is the ratio of the power reflected from a surface to the power incident on it. In Chapter 1 it was pointed out that R, for normal incidence, is related to the refractive index by

$$R = \left(\frac{n_1 - n_2}{n_1 + n_2}\right)^2 \tag{11–1}$$

where n_2 and n_1 are the refractive indices of the substrate material and incident medium, respectively. For conducting or absorbing dielectric materials, n_2 is a complex number and the square in Eq. (11–1) is an absolute square. The magnitude of the refractive index for good conductors (metals) is proportional to $\sqrt{\sigma/2\pi\mu f}$, where σ is electrical conductivity, μ is magnetic permeability, and f is the frequency of the light. Consequently, metals like copper and silver have high reflectances that increase with decreasing frequency (increasing wavelength). It is found that the reflectance of metals substantially decreases as the temperature nears the melting point.

2. Absorption coefficient α is the fractional loss of light power per unit distance for light traveling in a nonmetallic material. Beer's (Lambert's) law relates power to absorption by

$$P = P_0 e^{-\alpha z} \qquad (11\text{–}2)$$

where P_0 is the power entering the surface and P is the power at depth z. The absorption coefficient can also be interpreted as the penetration depth or distance at which the power has dropped to one over e (37%) of the value entering the surface.

3. Specific heat C is the energy required to raise the temperature of unit mass one degree. The SI units are joules per kilogram-Celsius degree but calories per gram-Celsius and BTU (British thermal unit) per pound—Fahrenheit degrees are also common. A frequently useful variation is the volume specific heat C_v given by ρC, where ρ is the mass density of the material. C_v is the energy required to raise the temperature of unit volume one degree, or energy per unit volume-degree.

4. Thermal conductivity k is the heat flow per unit area per unit thermal gradient. The units are presented in a variety of ways, but a look at the one-dimensional heat conduction equation, which relates rate of heat flow Q to thermal gradient dT/dz,

$$Q = -kA \frac{dT}{dz} \qquad (11\text{–}3)$$

indicates that the SI units for k are watt/m · °C, but watt/cm · °C are frequently used.

5. Thermal diffusivity κ is related to thermal conductivity and volume specific heat by

$$\kappa = \frac{k}{C_v} \quad = .0014\ cm^2/s\ (H_2 o) \qquad (11\text{–}4)$$

and is a measure of how much temperature rise will be caused by a pulse of heat applied to the material. It also indicates how rapidly heat will diffuse through the material. Units are generally cm²/s. Materials with high thermal diffusivity will experience a relatively small temperature rise with good heat penetration for a given heat pulse at the surface. A material with a low thermal diffusivity will undergo a relatively large temperature rise at the surface with low heat penetration into the material for a given heat pulse.

6. Latent heat L refers to the amount of heat required to cause a change of phase of unit mass of material. L_f is the latent heat of fusion or the energy required to cause melting of unit mass. Examples of units are cal/g, BTU/lb, and the SI units joule/kg. L_v is the latent heat of vaporization and has the same meaning and units as L_f. Latent heats of vaporization are much larger than latent heats of fusion and vaporization also requires more energy than that needed to raise unit mass up to the vaporization temperature. Consequently, when vaporization is involved in processing, it is a major factor in determining energy requirements.

7. Transformation temperatures refer to the melting temperature T_m, vaporization temperature T_v, and other phase change temperatures T_p, such as the martensitic transformation temperature on the iron-carbon phase diagram. The last is particularly important in heat treatment of cast iron and steel.

Laser beam parameters.   Laser parameters relate to both the laser beam and to the laser power output as a function of time. Normally lasers are either operated CW or pulsed; the pulses may be produced by Q-switching. The parameters of the beam that affect processing are

1. Wavelength
2. Focused spot size
3. Mode structure

Wavelength affects absorption and reflection characteristics and spot size and mode structure affect average irradiance and irradiance distribution in the spot.

The power output and traverse speed, or dwell time, are clearly important for CW operation. In pulsed lasers the length of the pulse, energy per pulse, peak power, and even the shape of the pulse influence the interaction of the beam and the workpiece.

Pulses from solid lasers usually consist of a substructure or spiking. The details of this spiking may have a significant influence on the quality of drilled holes.

11-2 BASIC HEAT TRANSFER EQUATIONS

The basic heat transfer equations considered are

$$\mathbf{F} = -k \ \nabla T \tag{11-5}$$

which relates heat flux $\mathbf{F}$ to thermal gradient T and

$$\nabla^2 T - \frac{1}{\kappa}\frac{\partial T}{\partial t} = \frac{-A}{k} \tag{11-6}$$

which applies to non-steady-state heat conduction with A accounting for internal generation of heat with units of watts/m³ or equivalent. A is zero in materials where the incident laser power is absorbed at the surface. This statement will be assumed for all remaining sections of this chapter except Section 11-8. When the laser beam penetrates the material to any extent, the solution of Eq. (11-6) requires consideration of a distribution of internal sources and $A(z, t)$ cannot be neglected.

Equation (11-5) is sufficient for steady-state heat transfer problems. Unfortunately, laser applications seldom lead to steady-state situations. For the purpose of this book, detailed solutions of Eq. (11-6) will not be undertaken, although solutions that pertain to laser applications will be presented and discussed. Generally radiation and convection are not significant sources of heat loss in laser processing and will be assumed negligible.

11-3 ENERGY BALANCE APPROXIMATION

Simple energy balance approximations frequently provide reasonable ballpark numbers for many applications. For that reason some discussion of this approach is presented here.

Figure 11-1 is a schematic representation of a laser beam focused onto the surface of a workpiece. If it is assumed that the material is heated to a depth z with cross-sectional area πa^2, then the energy U required is given by

$$U = (CT + L_f + L_v)\rho\pi a^2 z \tag{11-7}$$

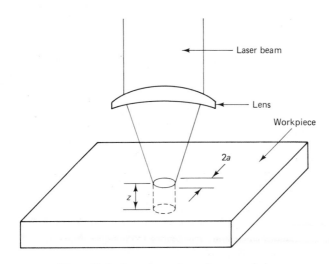

Figure 11-1 Laser beam focused on a workpiece.

where T is used for temperature to avoid confusion with time t and represents the change in temperature of the part. It has been assumed in Eq. (11–7) that ρ and C_v are independent of temperature and are the same for liquid and solid. Equation (11–7) includes the possibility of melting and vaporization. Only the first term is required for surface hardening without melting; the first two terms are used for welding; and all three are applicable for material removal, although the first two may be negligible. The depth of the cylinder heated in Fig. 11–1 may be taken as the thickness of the part, if complete penetration is desired, or to whatever depth requires heating. The depth is typically 0.25 to 1.3 mm in surface hardening.

Obviously the energy balance approach is crude; it ignores lateral heat transfer and assumes uniform heating throughout the cylinder. Nevertheless, it does provide a lower limit to the energy required in a given case and is fairly accurate for small z or $z < a$ and for t relatively short compared with the time required for significant heat loss by conduction.

Once U is determined, the length of the pulse, assuming constant power P, is just $t = U/P$. Looking ahead somewhat, the solution of Eq. (11–6) for constant uniform irradiance on the surface of a semi-infinite solid leads to a concept of thermal penetration depth, z_{th}, where

$$z_{th} = \sqrt{4\kappa t} \qquad (11\text{–}8)$$

Equation (11–8) can be used to estimate z in Fig. 11–1 for a given pulse length or, conversely, it can be used to estimate the pulse length required to achieve

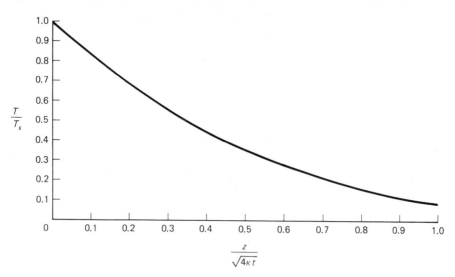

Figure 11–2   Ratio of temperature T at depth z to the surface temperature T_s versus $z\sqrt{4\kappa\tau}$ for uniform, constant irradiance.

a desired penetration depth. Actually, the temperature at $z = \sqrt{4\kappa t}$ is approximately one-tenth the temperature at the surface, relative to the initial temperature. Figure 11-2 is a plot of $T(z)/T_s$ against $\alpha = z/\sqrt{4\kappa t}$, where T_s is the temperature at the surface of the part. A depth of $z = \sqrt{\kappa t}$ is frequently used to estimate thermal penetration depth and is perhaps more reasonable, for at that depth the temperature change is about 35% of the surface temperature change.

The energy balance approach can be used with a moving source. Assume that the source in Fig. 11-1 is moving to the right with speed v (or the part is moving to the left with speed $-v$). The time that the beam remains on any one point is $t = 2a/v$. Then the energy delivered to a spot of radius a is $U = Pt$ or

$$U = \frac{2aP}{v} \tag{11-9}$$

Equation (11-9) can be used in conjunction with Eqs. (11-7) and (11-8) for problems involving a moving source or part. This treatment is quite good for nonmetals with high absorption coefficients.

11-4 UNIFORM, CONSTANT IRRADIANCE MODEL

A result of the uniform, constant irradiance model was presented in the previous section to improve the energy balance approach to laser processing problems. In this section the complete solution for the uniform, constant irradiance model is given.

In this model it is assumed that a semi-infinite solid is irradiated with a constant and uniform power per unit area. In all equations the irradiance I_0 at the surface is assumed to be the actual power per unit area absorbed, $(1 - R)I_i$, where I_i is the incident irradiance. The solution of Eq. (11-6) for the temperature, $T(z,t)$, as a function of depth z into the solid and time t is [Carslaw and Jaeger 1959]

$$T(z, t) = \frac{2I_0}{k} \left[\left(\frac{\kappa t}{\pi} \right)^{1/2} e^{(-z^2/4\kappa t)} - \frac{z}{2} \operatorname{erfc} \frac{z}{\sqrt{4\kappa t}} \right] \tag{11-10}$$

where

$$\operatorname{erf}(s) = \frac{2}{\sqrt{\pi}} \int_0^s e^{-x^2} \, dx$$

is the error function and erfc (s) is the complimentary error function or $1 - \operatorname{erf}(s)$. Tables of the error function are available in most compilations of mathematical tables, such as Jahnke and Emde, "Tables of Functions." The surface temperature of the semi-infinite solid is given by

$$T_s(0, t) = \frac{I_0}{k} \left(\frac{4\kappa t}{\pi} \right)^{1/2} \tag{11-11}$$

The graph in Fig. 11-2 was generated from Eqs. (11-10) and (11-11). Equation (11-11) is useful for estimating the time required to reach a specified surface temperature for a given irradiance.

In the case of a pulse of length t_p, $I(t) = I_0$, $T(z, t)$ is given by Eq. (11-10) for the duration of the pulse. For $t > t_p$ a solution is obtained by the principle of superposition, thus for $t > t_p$

$$T(z, t) = [f(t) - f(t - t_p)] \tag{11-12}$$

where $f(t)$ is used to represent the right-hand side of Eq. (11-10).

Equations (11-10) and (11-12) are valid when the thickness of the part exceeds $\sqrt{4\kappa t}$ and will give a reasonable estimate of the temperature along the axis of the beam extending into the part if the beam width exceeds $\sqrt{4\kappa t}$. Equations (11-10) and (11-12) are most accurately applied to laser heat treatment when a defocused beam is used, possibly in conjunction with beam scanning (dithering), rotation, or beam integration, to spread the power out in a more or less uniform fashion over a fairly large area.

11-5 GAUSSIAN, CIRCULAR, AND RECTANGULAR BEAM MODELS

Some pertinent results of models that more truly represent the nature of laser beams are given in this section. More thorough discussions of these models appear in several of the references, such as Duley 1976; Carslaw and Jaeger 1959; Ready 1971; and Charschan 1972. In all cases discussed in this section, the irradiance is constant in time, and the substrate is considered to be a semi-infinite region.

Gaussian beam.   Equation (11-6) can be solved for constant absorbed irradiance I distributed in a Gaussian fashion over the cross section of the beam with I_0 the absorbed irradiance at the center. Assuming that the power is absorbed at the surface, the surface temperature at the center of the spot is given by (11-7) [Duley 1976]

$$T = \frac{I_0 w}{k} \left(\frac{1}{2\pi} \right)^{1/2} \tan^{-1} \left(\frac{8\kappa t}{w^2} \right)^{1/2} \tag{11-13}$$

Circular beam.   The solution of Eq. (11-6) for the temperature at the center of a spot of radius a with power P uniformly distributed over the spot as a function of time and penetration depth z is [Duley 1976]

$$T(z, t) = \frac{2P(\kappa t)^{1/2}}{\pi a^2 k} \left[\text{ierfc} \frac{z}{(4\kappa t)^{1/2}} - \text{ierfc} \frac{(z^2 + a^2)^{1/2}}{(4\kappa t)^{1/2}} \right] \qquad (11\text{-}14)$$

where P is the total power absorbed at the surface and ierfc stands for the integral of the complementary error function and

$$\text{ierfc}(s) = \frac{1}{\sqrt{\pi}} e^{-s^2} - \text{serfc}(s)$$

The surface temperature at the center of the spot is given by

$$T(0, t) = \frac{P(4\kappa t)^{1/2}}{\pi a^2 k} \left[\frac{1}{(\pi)^{1/2}} - \frac{1}{(\pi)^{1/2}} \exp - \frac{a^2}{4\kappa t} \right.$$

$$\left. + \frac{a}{(4\kappa t)^{1/2}} \text{erfc} \frac{a}{(4\kappa t)^{1/2}} \right] \qquad (11\text{-}15)$$

The steady-state temperature as a function of penetration depth may be of interest. It is given by

$$T(z, \infty) = \frac{P}{\pi a^2 k} [(z^2 + a^2)^{1/2} - z] \qquad (11\text{-}16)$$

Therefore the maximum surface temperature that can be attained is

$$T_{\max} = \frac{P}{\pi a k} \qquad (11\text{-}17)$$

It is important to point out that the surface temperature reaches 75% of its steady-state value for $t = a^2/\kappa$, which for most metals and focused spots is a fairly short period of time. A focused spot of 0.025-cm radius and a diffusivity of 0.1 cm²/s, for example, gives $t = 6.4$ ms. The quantity a^2/κ is referred to as the thermal time constant and is a measure of the time required for significant radial heat conduction loss to occur.

The results for the circular and Gaussian spots do not differ by a great deal, particularly on the beam axis for values of $z > 0$. At the surface the steady-state temperature at the center of a Gaussian spot is about 25% greater than the temperature at the center of a circular spot for equal total power and $w = a$.

Rectangular beam. A laser operating in a higher-order mode may produce a beam whose cross section is approximately rectangular. If x and y are half the length and width of the spot, the steady-state maximum temperature is given by [Carslaw and Jaeger 1959]

$$T_{\max} = \frac{P}{2\pi xyk} \left(x \sinh^{-1} \frac{y}{x} + y \sinh^{-1} \frac{x}{y} \right) \qquad (11\text{-}18)$$

The maximum temperature for a square beam with $x = y = a$ is

$$T_{max} = \frac{0.885P}{\pi ak} \tag{11–19}$$

which is about 89% of the maximum steady-state temperature for a circular beam of radius a.

The constant irradiance model gives approximately the same results as the three-dimensional models for a dwell time less than the thermal time constant a^2/κ, which is, as has been shown, typically in the range of milliseconds. The uniform irradiance model can provide order of magnitude estimates for milli-second-range welding pulses and for continuous operation if the travel rate is high. A speed of at least 80 mm/s (189 in./min) for steel would be required for a spot diameter of 0.5 mm in order to satisfy the criterion that the beam dwell time on any one spot be less than the thermal time constant. If this criterion is not met, one of the other models should be used.

11–6 HEATING WITH MELTING

The process of melting a material with the energy supplied by the laser is relatively simple to analyze if the time to bring the surface to the melting point is short compared with the thermal time constant. The time to reach melting at the surface t_m can be estimated from the uniform irradiance model by

$$t_m = \frac{\pi}{4\kappa}\left(\frac{kT_m}{I_0}\right)^2 \tag{11–20}$$

This is usually much less than a microsecond for typical welding situations. On the basis of energy balance, the volume of material melted can be estimated as

$$V = \frac{U}{\rho(CT_m + L_v)} \tag{11–21}$$

where U is the total energy input. Because conduction losses have been neglected, U is a lower limit to the energy required. Doubling U usually gives a reasonable estimate of the actual energy needed. The length of time to reach vaporization can be estimated from

$$t_v = \frac{\pi}{4\kappa}\left(\frac{kT_v}{I_0}\right)^2 \tag{11–22}$$

It is typically milliseconds for pulsed welding with I_0 on the order of 10^5 W/cm². The parameters κ and k are not constant with temperature, but reasonable calculations can be made by selecting an intermediate value for the range of temperatures under consideration. Because more time is spent in the molten state than the solid state during heating, the most appropriate values of κ and k are averages over their values in the molten range of interest.

As an example, consider the welding of steel with a CO_2 laser capable of providing a power of 2.5K W at the workpiece. A weld depth of 2 mm is desired with a weld width of about 1 mm. The appropriate parameters are

$$T_m = 1547°C \qquad T_v = 2752°C$$

Room temperature is $T_a + 16°C$, $\rho = 7.87$ g/cm³, $L_f = 272$ J/g, $\kappa = 0.21$ cm²/s, $k = 0.75$ W/cm°C, and $C = 0.46$ J/g°C.

According to energy balance, $U = \rho V[CT_m + L_f] = 11$ J. U should be doubled to compensate for heat conduction losses. Assuming that half the power is coupled into the part, $t_p = U/P = 19$ ms. In actual practice, a high-order CO_2 laser beam focused to a spot diameter of approximately 0.5 mm was used with a traverse speed of 2.54 cm/s. The dwell time of this beam on any one spot is $t = 0.05$ cm/(2.54 cm/s) $= 20$ ms. The close agreement of these times is coincidental, for the choice of a 1.0-mm wide weld and 50% reflection are somewhat arbitrary. According to the uniform irradiance model, the time to reach T_v at the surface is

$$t_v = \frac{\pi}{4\kappa}\left(\frac{kT_v}{I}\right)^2 = 0.04 \text{ ms}$$

Obviously this time is much too short to allow any useful penetration without substantial material removal occurring. The development of a "keyhole"[1] in the high-power continuous welding, which allows the beam to penetrate unimpeded deep into the part, is absolutely essential and is what makes possible a reasonable estimate using the energy balance approach. It is believed, however, that nearly all the incident energy is coupled into the part during keyholing. An irradiance of 5×10^5 W/cm² (the irradiance used in this example) will tend to drill except in the case of continuous welding where a keyhole develops.

The uniform irradiance solution gives reasonable results in microwelding of small, thin parts. If the penetration depth l and I_0 are known, the graph in Fig. 11–2 can be used to find a value of α at $T(z)/T_s = T_m/t_v$. Then the pulse length is obtained from $\alpha = l/\sqrt{4\kappa t_p}$.

Another approach, which takes into account the formation of a liquid front and its propagation through the workpiece, has been developed by Cohen and Epperson, 1968. Uniform, constant irradiation for some time t is assumed in this approach; therefore good results may be expected when the spot size exceeds the penetration depth and when $t < a^2/\kappa$, where a is a dimension characterizing the spot size, such as the radius for a circular spot. The solution was obtained via an analog computer and is expressed in graphical form in Fig. 11–3. To use this graph, calculate $T = 40T_v/T_m$ and $Y = L_f/CT_m$ (temperatures are relative to ambient). Locate T on the left-hand vertical scale and go

---

[1] The phenomenon of "keyholing" is discussed in more detail in Section 11–9.

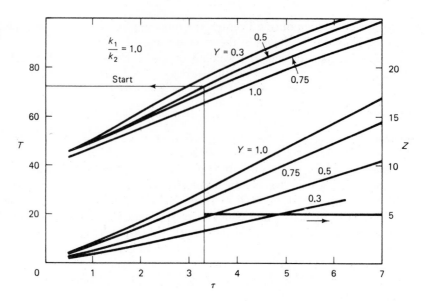

Figure 11–3 Cohen and Epperson's technique for determining melt depth. $T = 40$ T_v/T_m, $Z = 8.33 \, \rho L_f z/It_m$, $\tau = t_v/t_m$, $Y = L_f/CT_m$. (Adapted with permission of the *Journal of the Franklin Institute*, Vol. 283, No. 4, April 1967, pp. 271–85.)

horizontally across to the appropriate upper curve for Y. Drop vertically to the corresponding Y curve of the lower set. Dropping vertically to the horizontal axis from this last point provides $\tau = t_v/t_m$ and proceeding horizontally to the right vertical axis gives $Z = 8.33 \, \rho L_f z/It_m$. The times t_v and t_m are the times to reach vaporization and melting at the surface, respectively; t_m is calculated from Eq. (11–1). The depth of penetration z is determined from the expression for the nondimensional depth Z.

An example of the application of Cohen and Epperson's technique is obtained by considering spot welding of two sheets of 0.4-mm-thick steel together with a beam focused to a spot diameter of 1.25 mm. The most suitable laser for this application would be a Nd-YAG or Nd-Glass laser. In this application it is assumed that the pulse length and absorbed irradiance are unknown. The penetration depth z equals 0.8 mm and the surface temperature T_s equals T_v at the end of the pulse. Using the parameters given for the previous example and Eq. (11–11), we have

$$I = \frac{\pi k^2 T_m^2(z)}{(4)\rho 8.33 \kappa L_f Z} = 13 \times 10^3 \text{ W/cm}^2$$

which means a total power coupled into the piece of $I\pi(0.063 \text{ cm})^2 = 160$ W. Assuming 50% reflectance, 320 W would be required. The pulse length is

$$t_m = \frac{\pi}{4(0.21)}\left[\frac{0.75(1526)}{13 \times 10^3}\right]^2 = 29 \text{ ms} \qquad t_p = t_v = \tau t_m = 87 \text{ ms}$$

This time is considerably longer than the thermal time constant, which is $(0.063$ cm$)^2/(0.21$ cm^2/s$) = 19$ ms. Consequently, a fairly large quantity of heat will be lost to conduction. Much better results are obtained when the weld depth is much less than the spot diameter. A higher irradiance can be used for thinner materials, leading to much shorter times to reach vaporization with full melting through the piece. One point that this example makes is that there is a limit to weld depth (without keyholing) unless significant material removal is allowed at the surface. In practice, weld depths in laser or electron beam welding would be limited to about 1 to 2 mm if it were not for keyholing.

11-7 HEATING WITH VAPORIZATION

At power densities sufficient to produce rapid vaporization, typically 10^6 W/cm^2 and higher, the time to reach vaporization can be estimated from Eq. (11–10) because very little heat penetration occurs in this length of time. Using the data for steel and an irradiance of 10^6 W/cm^2, for example

$$t_v = \frac{\pi}{4\kappa}\left(\frac{kT_v}{I}\right)^2 = 16 \text{ μs}$$

Once vaporization occurs, a vapor front begins to move into the material, preceded by a liquid front. This situation is depicted in Fig. 11–4. If the material removal process occurs in a relatively short period of time, the energy requirement and pulse time can be reasonably estimated by using the energy balance approach. The energy required to remove a given volume of material is given by Eq. (11–7). For a circular spot of radius a and thickness of material removed

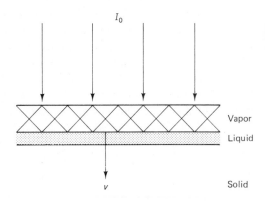

Figure 11-4   Vapor front propagation in a solid.

z, $V = \pi a^2 z$. It has been assumed that the specific heat and density are constant with respect to temperature and are the same for solid and liquid. The heat of fusion has been neglected. Letting the depth of material removed be a variable and taking the derivative of Eq. (11-7) yield

$$P = \frac{dU}{dt} = \rho \pi a^2 v (CT_v + L_v) \tag{11-23}$$

where $v = dz/dt$ is the speed at which the vapor front moves into the material. The liquid interface between the vapor and solid is extremely thin and is ignored in this analysis. Because $I = P/\pi a^2$, the speed of the vapor front can be calculated from

$$v = \frac{I}{\rho(cT_v + L_v)} \tag{11-24}$$

The time t_p required to remove material to a depth z, is then simply

$$t_p = \frac{z}{v} \tag{11-25}$$

As an example, consider piercing a 0.5-mm-diameter hole through a 1-mm-thick sheet of steel with $L_v = 6350$ J/g. The energy required, using Eq. (11-21), is

$$U = 11.8 \text{ J}$$

If the irradiance is 10^6 W/cm^2, the pulse length is, by Eqs. (11-24) and (11-25),

$$t_p = 6 \text{ ms}$$

Actual drilling times are 0.1 to 1.0 ms in practice.

If instead of piercing a hole it is assumed that the material is being cut, in the previous example, the cutting rate can be estimated from $v = 2a/t_p = 0.05$ cm/6 ms $= 8.3$ cm/s (197 in./min). This is in surprisingly good agreement with experience. Oxygen assist is used to oxidize the metal and to blow away molten material in most metal-cutting applications. In this technique 60 to 70% of the cutting energy is supplied by the oxidation process and not all the material removed from the kerf is vaporized. The energy balance approach assumes that all the kerf material is vaporized and that no conduction losses occur. These factors are somewhat compensating, which helps to account for the accuracy of the crude energy balance calculation.

When material removal involves pulses with variable power, the energy balance approach is still relatively easy to use if the temporal dependence of the power is known. Using Eq. (11-24), the depth of material removed can be expressed as an integral

$$z = \frac{1}{\rho(CT_v + L_v)} \int_0^t I(t)\, dt \tag{11-26}$$

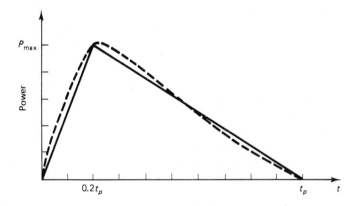

Figure 11-5 Generalized laser pulse. (Adapted from J. F. Ready, *Effects of High-Power Laser Radiation*, with permission of Academic Press.)

This integral can be evaluated for a generalized pulse,[2] which applies to a large number of pulsed or Q-switched lasers. Figure 11-5 is a plot of this generalized pulse.

The irradiance is given by $I_1(t) = 5I_{max}(t/t_p)$ up to $t = 0.2t_p$. And from $t = 0.2t_p$ to $t = t_p$ the irradiance is given by $I_2(t) = 1.25\ I_{max}\ (1 - t/t_p)$. Putting these expressions into Eq. (11-25) and integrating yield

$$l = \frac{0.5I_{max}}{\rho(CT_v + L_v)} \tag{11-27}$$

Of course, this result, for a simple triangular pulse, could have been written down immediately, based on the area under the curve in Fig. 11-5. Inherent in this approach is the assumption that the vaporization temperature at the surface is reached in a time much less than t_p and that full penetration is reached before I drops too low to maintain a vapor front.

The rate of material removal for irradiances of the order of 10^8 W/cm^2 becomes limited by the rate at which atoms can leave the surface of the vapor front. The ultimate limiting velocity of atoms in the vapor is the speed of sound in the solid—that is, the maximum speed with which an atom can escape. If the rate of energy input is too great, the vapor does not have time to clear the way for the laser beam and the vapor becomes superheated—that is, $T > T_v$. Consequently, much of the input energy ends up being absorbed by the vapor. In fact, short, high peak power pulses tend to give rise to a vapor plume that rises above the workpiece and may totally attenuate the laser beam, thereby wasting most of the energy in the pulse. Longer, lower peak power pulses are, in fact, superior for most hole-piercing applications. Such pulses are typically of the order of magnitude of 0.1 to 1.0 ms and the energy balance approach

[2] Suggested by Ready, 1971, p. 75.

is reasonable. Because most industrial applications are of this nature, the energy balance approach is the only approach presented here. More rigorous treatments are available in more advanced books [Duley 1976, Charschan, et al. 1977].

11-8 DISTRIBUTED SOURCES

A brief discussion of the results of the analysis of internal absorption of a laser beam is presented in this section.

The basic law governing the absorption of light in a dielectric medium in Beer's (Lambert's) law.

$$I = I_0 e^{-\alpha z} \tag{11-28}$$

where I_0 is the irradiance entering the surface of the medium at $z = 0$ and α, the absorption coefficient, is the fractional loss of irradiance per unit length. The reciprocal of α is the depth at which the irradiance has dropped to $1/e$ (37%) of its value entering the surface and can be thought of as a measure of penetration depth. The absorption coefficient for many plastics, such as polycarbonate, is 0.33 cm^{-1} for visible light. ZnSe has an absorption coefficient of 1.0×10^{-3} cm^{-1} at a wavelength of 10.6 μm. Glass has an absorption coefficient of 0.05 cm^{-1} for visible radiation. Plastics have an absorption coefficient in the range of 10 to 20 cm^{-1} at a wavelength of 10.6 μm.

$(0.10cm^{-1})$

For uniform irradiance, constant in time, the solution of the heat conduction equation, including a distributed source, leads to the result that the temperature of the heated material rapidly rises to a value given by [Duley 1976]

$$T = \frac{0.6 I}{k \alpha} \tag{11-29}$$

to a depth of approximately α^{-1}. The thermal conductivities of plastics range from 0.001 to 0.004 W/cm°C. This means that less than 100 W/cm^2 is required to raise plastics to their melting point to a depth of 0.5 mm almost instantly when using a CO$_2$ laser. $6mW/cm°C$ $[H_2O]$

11-9 CW WELDING (KEYHOLING)

With the advent of CO$_2$ lasers with 1.0 kW and higher CW power output, it was discovered that the phenomenon of keyholing occurs for continuous seam welding. A similar phenomenon was known to occur for multikilowatt electron beam welding. In this phenomenon a hole is produced in the material, allowing the beam to penetrate relatively unimpeded into the part. Apparently at power inputs of 1 kW and higher the vapor pressure of the molten material becomes sufficiently high to overcome surface tension and pushes the molten material out of the way, forming a hole, or cavity, that is virtually 100% absorptive. The molten material flows back into the hole after the beam has passed. Generally the higher the power, the deeper is the cavity that can be formed. This phenomenon is illustrated in Fig. 11–6.

. If it were not for the formation of a keyhole, laser (as well as electron beam) welding would be limited to penetration depths of about 1 mm by the thermal properties of metals. Using keyholing, weld depths of several centimeters have been achieved with very high power CW CO_2 lasers.

Swift-Hook and Gick have modeled the keyholing process by use of a linear heat source of total power P extending into the metal a distance a [Swift-Hook and Gick 1973]. Referring to the geometry depicted in Fig. 11–7, their result for the temperature distribution is

$$T = \frac{P}{2\pi ak} \, exp \left(\frac{vx}{2\kappa} \right) K_0 \left(\frac{vr}{2\kappa} \right) \qquad (11\text{–}30)$$

Here v is the weld speed, K_0 is the Bessel function of order zero, and $r = \sqrt{x^2 + y^2} = x \cos \phi$. The width of the weld can be determined by letting $T = T_m$, $\phi = 90°$ and solving for $r = w/2$. For high welding speeds, Eq. (11–31) gives a reasonable estimate of weld width.

$$w = 0.484 \, \frac{\kappa}{v} \frac{P}{akT_m} \qquad (11\text{–}31)$$

This will apply to most multikilowatt CO_2 laser welding. The graph on p. 198 of Duley (11–7) can be used to determine whether Eq. (11–31) is applicable.

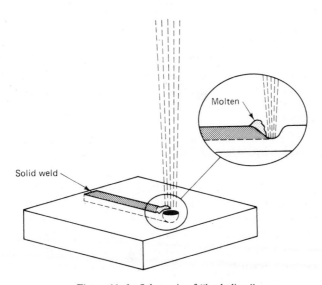

Figure 11–6 Schematic of "keyholing."

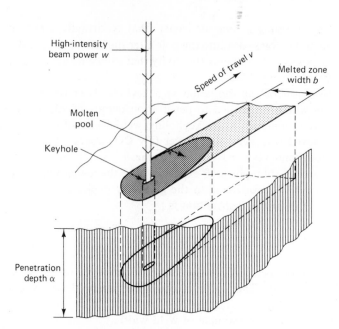

Figure 11–7   Linear heat source model of laser keyholing. (Courtesy of *Optical Engineering* and United Technologies Research Center.)

11–10 GENERAL APPROACH TO LASER HEATING

Many treatments of the general problem of heating a solid with a laser exist. Considerable success has been achieved if surface absorption of the beam is a valid assumption, particularly in surface modification applications in which melting and vaporization are not involved. Complete solutions in closed form are rarely, if ever, possible for the real cases. Numerical, finite difference and finite element techniques have been used with success, however. Several references are listed at the end of the chapter for the reader interested in pursuing this subject at some length. Most solutions include the assumption of constant thermal constants, which permits the use of the principle of superposition, based on the linear nature (with this assumption) of the general heat conduction equation (Eq. 11–2).

Difficulties are involved in the application of any of these approaches. The variation of thermal constants with temperature is not well known for materials of interest. An absorptive coating must be used in heat treatment with CO_2 lasers and the absorption efficiency (ratio of absorbed power to incident power) and its variation with temperature and coating thickness are not well understood. To complicate matters, the measurement of laser power is only fairly accurate for long-wavelength lasers, such as that of the CO_2 laser. Nevertheless, the qualitative agreement between calculated and measured temperature profiles is excellent and the quantitative agreement is quite good.

An approach that is straightforward and that provides considerable insight into heating with a laser will be described in some detail. It is a linear approach for which the solution of Eq. (11–2) for an instantaneous point source serves as a basis. This solution is

$$T(x, y, z, t) = \frac{U}{4\rho C(\rho \kappa t)^{3/2}} \exp - \frac{[(z - x')^2 + (y - y')^2 + z^2]}{4\kappa t} \tag{11–32}$$

where U is the energy liberated at time $t = 0$ at point $(x', y', 0)$ on the surface of a semi-infinite medium [Charschan 1972]. Using the principle of superposition allows solutions for any type of spatial and temporal energy distribution to be

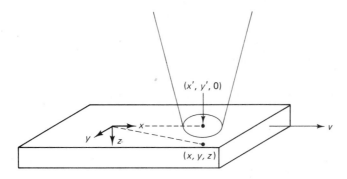

Figure 11–8   Defocused laser beam incident on a part.

obtained. Several solutions have been reported in the literature. Arata et al. have given results for moving rectangular, circular, and Gaussian heat sources and have compared these results with experimental findings and with the predictions of the uniform irradiance model [Arata, Marou, Miyamoto 1978]. The moving Gaussian heat source is the easiest to solve and provides the simplest result. The results for the Gaussian source will be presented here to illustrate the approach and to provide understanding of the laser heating process. The Gaussian model is the least satisfactory of the three for actual surface modification applications, but the results for the three models differ only slightly for predicted temperatures near the center line of the laser beam.

Figure 11–8 depicts a Gaussian laser beam incident on a part moving in the x direction with speed v. It is assumed that the instantaneous energy released is given by $U = I(x', y') \, dx' \, dy' \, dt'$, where for a Gaussian distribution

$$I(x', y') = I_0 e \exp - [x'^2/a^2 - y'^2/b^2]$$

To account for the fact that the part is moving, x' must be replaced by $x - v(t - t')$. The temperature distribution for a point source is thus given by

$$T(x,y,z,t) = \frac{I_0}{4\rho C[\pi\kappa(t - t')]^{3/2}}\left[\exp -\frac{[x - x' + v(t - t')]^2}{4\kappa(t - t')}\right.$$

$$\left. -\frac{x'^2}{a^2} - \frac{(y - y')^2}{4\kappa(t - t')} - \frac{y'^2}{b^2} - \frac{z^2}{4\kappa(t - t')}\right] dx'\, dy'\, dt' \qquad (11\text{--}33)$$

In Eq. (11–33) t has been replaced by $(t - t')$ because the latter represents the time between the instant that the heat was released and the time of interest, t; x, y, z are the coordinates of the point at which the temperature value is desired whereas $x' - v(t - t')$ and y' give the location of the heat source at time t'. Equation (11–33) must be integrated for a continuous Gaussian source. The limits on the x and y integrals are, because of the continuity of the Gaussian function, $-\infty$ to $+\infty$. The time integral is taken from $-\infty$ to the present time $t = 0$. For any fixed point in space, the temperature rapidly approaches a steady-state value, which is clearly the maximum value of the temperature at that point. It is the maximum temperature isotherms that are of greatest interest in surface modification applications.

If the variables are changed, the time integration reduces to an integration over t' from zero to ∞. By completing the squares in the exponentials, the integrand can be put in a form that is integrable over the spatial coordinates. Then after some algebra the steady-state temperature distribution for a Gaussian heat source is

$$T(x, y, z) = \frac{Pk}{(\pi\kappa)^{3/2}}\int_0^\infty \frac{dt'}{(t')^{1/2}} \times$$

$$\frac{\exp\left(-\dfrac{(x + vt')^2}{a^2 + 4\kappa t'} - \dfrac{y^2}{b^2 + 4\kappa t'} - \dfrac{z^2}{4\kappa t'}\right)}{\sqrt{(b^2 + 4\kappa t')(a^2 + 4\kappa t')}} \qquad (11\text{--}34)$$

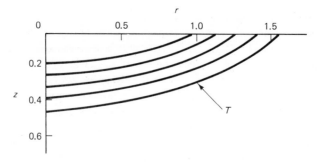

Figure 11–9 Isothermal lines for a Gaussian beam heated part. z, r, and T are normalized values. (Adapted with permission of Prof., Dr. Yoshiaki Arata, from *Application of Laser for Material Processing.*)

where $P = I_0(\pi ab)$ is the total power entering the part. The integration in Eq. (11–34) can be carried out numerically to determine the temperature isotherms. Arata et al. present the isotherms in nondimensional form so that they can be applied to any case regardless of the power, parameter values, or spot size (see Fig. 11–9) [Arata, Marou, Miyamoto 1978].

PROBLEMS

11–1. (a) Using the energy balance approach, estimate the power required to produce a spot weld 3 mm in diameter and 1 mm deep in low-carbon steel with a 1-s pulse from a CO_2 laser.

$$\rho = 7.86 \text{ g/cm}^3, \, \kappa = 0.07 \text{ cm/s}$$
$$C = .7 \text{J/g°K}, \quad k = 0.4 \text{ W/cm°K}$$

Estimated values near melting point are

$$T_m = 1809°\text{K}$$
$$L_f = 13.8 \text{ kJ/mol}$$

Assume 60% reflectance.

(b) Compare the depth given in part (a) with the thermal penetration depth predicted on the basis of the uniform irradiance model. Remember, at this depth the temperature is only about 35% of the surface temperature.

(c) According to the uniform irradiance model, how long does it take to reach T_m at the surface for the power deduced in part (a)?

11–2. Repeat Problem 11–1(a), using the uniform irradiance model and calculate the temperature at the surface at $t = 1$ s.

11–3. Apply the technique of Cohen and Epperson to Problem 11–1(a) to find the necessary power. $T_v = 3135°\text{K}$, $L_v = 350$ kJ/mole.

11–4. Compare the surface temperatures at the center of the spot for Gaussian and circular beams of equal radii of 0.5 mm ($1/e^2$ point for the Gaussian beam) incident on titanium if the total power in each beam is 500 W and the pulse length is 1.5 ms. Use the following data for titanium.

$$k = 0.219 \text{ W/cm°K}$$
$$\kappa = 0.1 \text{ cm}^2/\text{s}$$
$$50\% \text{ reflectance}$$

11–5. Calculate the maximum attainable surface temperature for an 8- by 6-mm rectangular beam for 2.5 kW incident on steel.

$$k = 0.8 \text{ W/cm°K}$$
$$60\% \text{ reflectance}$$

11–6. Low-carbon steel of 5-mm thickness is to be cut with a laser at a speed of 1.25 m/min with a 0.5 mm kerf. Using energy balance, estimate the power that must be coupled into the material.

$$k = 0.4 \text{ W/cm}°\text{K}$$
$$\rho = 7.86 \text{ g/cm}^3$$
$$C = 0.50 \text{ J/g}°\text{K}$$
$$L_v = 3.50 \text{ kJ/mol}$$
$$L_f = 13.8 \text{ kJ/mol}$$
$$T_v = 3135°\text{K}$$

11-7. Mild steel is to be welded with a multikilowatt CO_2 laser. Using the model of Swift-Hook and Gick, estimate the power required to produce a weld of 2 mm width and 4.5 mm depth at 1 m/min.

$$\kappa = 0.09 \text{ cm}^2/\text{s}$$
$$k = 0.4 \text{ W/cm}°\text{K}$$
$$T_m = 1809°\text{K}$$

11-8. A plastic material is irradiated with a CO_2 laser beam with a uniform irradiance of 20 W/cm^2. Calculate the instantaneous temperature rise.

$$k = 0.003 \text{ W/cm}°\text{K}$$
$$\alpha = 20 \text{ cm}^{-1}$$

11-9. A 0.4-mm diameter copper wire is to be laser welded to the top of a 1-mm diameter, 70 to 30% brass post. For a focused spot size of 0.4 mm and a melt depth into the post of 0.5 mm, estimate the required energy and pulse length. Assume 50% reflectance.
The properties of Cu and brass are as follows:

Cu	Brass (70–30)
$\rho = 8.96 \text{ g/cm}^3$	$\rho = 8.53 \text{ g/cm}^3$
$C = 0.38 \text{ J/g}°\text{K}$	$C = 0.37 \text{ J/g}°\text{K}$
$T_m = 1358°\text{K}$	$T_m = 1230°\text{K}$
$L_f = 255 \text{ J/g}$	$L_f = 200 \text{ J/g}$

11-10. Holes are to be drilled in 0.5-mm-thick nickel with a diameter of 0.13 mm, using a pulsed ND-YAG laser with a 4 MW peak power output. Use energy balance to estimate the energy needed and the pulse length, assuming a nearly square pulse. Show that the time to reach vaporization at the surface is negligible, based on the results of the uniform irradiance model.

$$k = 0.91 \text{ W/cm}°\text{K}$$
$$C = 0.44 \text{ J/g}°\text{K}$$
$$\rho = 8.9 \text{ g/cm}^3$$
$$L_f = 298 \text{ J/g}$$
$$L_v = 6303 \text{ J/g}$$
$$T_m = 1726°\text{K}$$
$$T_v = 3187°\text{K}$$
$$R = 60\%$$

REFERENCES

ARATA, Y., H. MAROU, and I. MIYAMOTO, "Application of Laser for Material Processing—Heat Flow in Laser Hardening," *IIW Doc. IV-241–78. Japan: Osaka University, 1978.*

CARSLAW, H. S., and J. C. JAEGER, *Conduction of Heat in Solids* (2nd ed.). New York: Oxford University Press, 1959.

CHARSCHAN, S. S. (ed.), *Lasers in Industry.* New York: Van Nostrand Reinhold Co., 1972.

CHARSCHAN, S. S., et al. (eds.), *The LIA-Material Processing Guide*, Toledo, OH: Laser Institute of America, 1977.

COHEN, M. I., and J. P. Epperson, *Electron Beam and Laser Beam Technology.* New York: Academic Press, 1968.

DULEY, W. W., *CO_2 Lasers: Effects and Applications.* New York: Academic Press, 1976.

DULEY, W. W., et al., "Coupling Coefficient for CW CO_2 Laser Radiation on Stainless Steel," *Optics and Laser Technology*, December 1979.

READY, J. F., *Effects of High-Power Laser Radiation.* New York: Academic Press, 1971.

SWIFT-HOOK, D. T., and A. E. F. GICK, "Penetration Welding with Lasers," *Welding Research Supplement*, pp. 492–499, November 1973.

BIBLIOGRAPHY

11–1 Ferris, S. D., H. J. Leamy, and J. M. Poate (eds.), "Laser-Solid Interactions and Laser Processing-1978," *AIP Conference Proceedings No. 50.* New York: American Institute of Physics, 1979.

11–2 Duley, W. W., *Laser Processing and Analysis of Materials.* New York: Plenum Press, 1983.

11–3 Bertolotti, M. (ed.), *Physical Processes in Laser-Materials Interactions.* New York: Plenum Press, 1983.

11–4 Bass, M. (ed.), *Laser Materials Processing.* New York: North-Holland, 1983.

chapter 12

High-Power Laser Applications

The purpose of this chapter is to discuss the techniques used in materials processing with lasers and to point out some of the advantages and disadvantages. The primary concern is with heat treatment, welding, cutting, hole piercing, scribing, and marking. A brief discussion of laser-assisted machining (LAM), alloying, and cladding is included. Appropriate theoretical techniques are pointed out and some quantitative information is presented to provide the reader with a feel for laser processing. There is, however, no intent to provide a compilation of laser applications and processing data. Reference may be made to many excellent books of that nature.

12–1 SURFACE HARDENING

Laser surface hardening (heat treatment) is a process whereby a defocused beam (generally from a 1.0 kW CW or higher-power CO_2 laser) is scanned across a hardenable material to raise the temperature near the surface above the transformation temperature. Normally the cooling rate due to self-quenching by heat conduction into the bulk material is sufficiently high to guarantee hardening. Steel with over 0.2% carbon and pearlitic cast irons are readily hardened by this technique.

There are several potential advantages to laser surface hardening:

1. The low amount of total energy input to the part produces a minimal amount of distortion and frequently parts can be used "as is" after laser hardening.

2. The depth of hardening is relatively shallow; thus desirable properties of the substrate metal, such as toughness, are retained.

3. Selective areas can be hardened without affecting the surrounding area.

4. In some cases, the desired wear characteristics can be achieved in a lower-cost material.

The chief disadvantages to laser hardening are

1. Low-area coverage rate, which limits the process to low production rates or selective hardening.

2. The high reflectance of metals of 10.6-μm radiation means that absorptive coatings must be used.

Hardening with a laser beam does not differ fundamentally from any other hardening process. In steel or iron the temperature at the surface is raised above the Martensitic transformation temperature, but is maintained below the melting point, as the beam scans the surface. The cooling rate near the surface can easily exceed $1000°C/s$; so the result is a hardened zone down to a depth at which the temperature exceeded the lower Martensitic transformation temperature. Figure 12–1 is an illustration of this process. A defocused beam of approximate size d produces a hardened track of approximately the same width and a hardened depth t determined by the laser power and coverage rate.

Most CO_2 laser beams are not suitable for heat treating as they come from the laser due to irregularities in the power density (irradiance) in the beam cross section. Gaussian and other lower-order modes are not acceptable because of the highly peaked irradiance distribution even if irregularities are not present. Single higher-order mode beams are especially suited to hardening

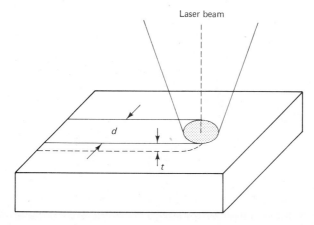

Laser beam

Figure 12–1 Defocused laser beam incident on workpiece for heat treating.

because of the large irradiance near the edges of the beam; this factor tends to compensate for lateral heat loss and produces a more nearly rectangular heat-treat-track cross section. Unfortunately, it is difficult to maintain a high-quality, high-order mode over a long period of time. Also, the beam may not be symmetrical, which complicates matters if contour tracks are required.

Some method to produce a nearly uniform average irradiance profile over a specified spot area is used in many production applications. It may be achieved by rotating the beam optically, thereby producing an overlapping spiral track or by dithering the beam (rocking the lens or a mirror) perpendicular to the track, thus producing a zigzag pattern. Figure 12–2 is an illustration of these techniques. Other techniques, referred to as beam integration, use stationary optical devices to produce relatively uniform beam images or spots. Some are basically hollow, polished copper tubes or pipes. The irregular beam enters one end and a fairly uniform beam exits at the other. Another device, available from Spawr Optical Company, is a segmented mirror. Each of about 32 segments independently forms a rectangular image of the light that strikes it. The super-position of the images formed by the irradiated segments provides a remarkably uniform rectangular spot in the image plane. Spot size can be varied by the use of additional optics. Figure 12–3 is a photograph of a Spawr segmented mirror.

Actual heat-treat-track profiles agree quite well with calculated ones. Numerical integration, finite difference, and finite element analysis techniques have all been successful in predicting heat-treat-track profiles for various irradiance distributions (see Chapter 11). Depth of penetration can be reasonably well predicted on the basis of the uniform irradiance model, which has an error function solution. The chief difficulty with any theoretical approach is the uncer-

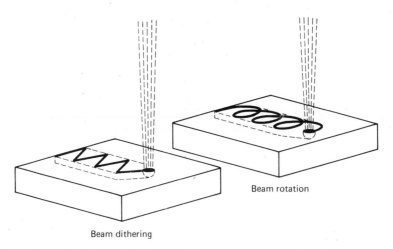

Beam rotation

Beam dithering

Figure 12–2 Active methods for spreading out a laser beam for surface treatment applications.

Figure 12–3 Segmented mirror for beam integration. (Courtesy of Spawr Optical Research, Inc.)

tainty in the value of the absorption coefficient[1] and the thermal parameters of the material.

Commonly used absorption coatings are black spray paint, graphite, phosphate, and zinc phosphate. It appears that paint is most effective at the higher-power levels (~5 kW) because a fairly thick coating is required. Manganese phosphate (Lubrite) is a common industrial coating and serves as a suitable absorptive coating for lower-power levels of around 1 kW. Zinc phosphate has comparable properties to manganese phosphate and graphite sprays work well but are generally expensive.

The absorption coefficient for uncoated steel or iron may be as low as 10 to 40%, depending on surface condition. A suitable coating raises the absorption coefficient to 60 to 80%.

It is important that the proper thickness of coating be used. If the coating is too thick, too much energy is wasted ablating the coating. If the coating is too thin, it completely burns off before the beam has passed and the bare metal reflectance occurs for part of the pass. What apparently happens when the coating thickness is correct is that heat is absorbed by the coating and transferred to the metal by thermal conduction at a rate that prevents the coating from being totally destroyed. In effect, the metal provides a heat sink for the energy entering the coating and the coating is removed at a rate that allows maximum coupling of beam energy into the part.

Figure 12–4 contains micrographs of heat treat tracks in 1045 steel and pearlitic cast iron. Hardness in the Martensitic region is near the maximum predicted by the carbon content. A thin overtempered zone may exist at the border of the hardened area. The ideal cross section for a heat treat track is a rectangle to minimize overlapping, variation in hardened depth, and the amount of overtempered zone at the surface.

---

[1] Ratio of absorbed power to incident power.

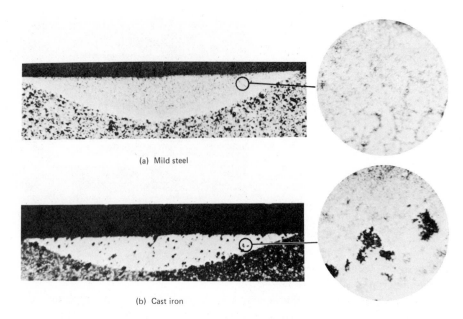

(a) Mild steel

(b) Cast iron

Figure 12–4 Micrographs of heat treat tracks. (Courtesy of Saginaw Steering Gear, Division of GMC.)

(a)   (b)

Figure 12–5 Laser-heat-treated parts. [(a) Courtesy of Saginaw Steering Gear, Division of GMC. (b) Courtesy Electro-Motive Division of GMC.]

Ideally an inert cover gas should be used during laser heat treatment to minimize oxidation of the metal and flame formation from the coating. It is common practice, however, to use air for economic reasons. A strong cross flow should be used to protect the final optics from smoke and spatter from the coating.

Figure 12–5 contains photographs of production laser heat-treated parts. Figure 12–6 is a photograph of an industrial heat treatment setup. Figures 12–7 and 12–8 contain plots of penetration depth versus coverage rate for two different materials.

Optimizing penetration depth in terms of coverage rate is a tradeoff between rapid and slow heating. If the surface is heated too rapidly, little heat enters the metal before melting occurs and a shallow depth is obtained. If the coverage rate is too slow, a substantial amount of heat is lost by thermal conduction. Generally higher power is required to achieve greater hardened depth. To some extent, a narrower track can be run to increase power density, but narrow tracks result in a significant amount of heat being lost by lateral (parallel to the surface) heat conduction. Typical hardened depths in mild steel are 0.5 mm at 1 kW, 1.0 mm at 2.5 kW, and 2.0 mm at 5.0 kW.

Figure 12–6 Industrial heat treating setup. (Courtesy of Spectra-Physics, Industrial Laser Division.)

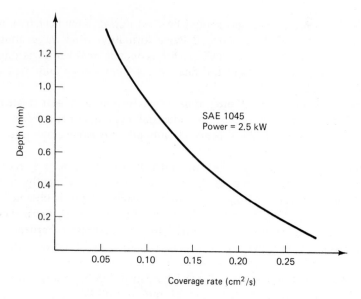

Figure 12-7 Hardened depth versus coverage rate for SAE 1045 for a 0.12-cm diameter beam.

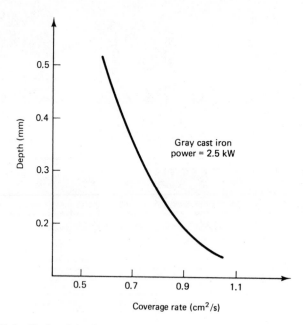

Figure 12-8 Hardened depth versus coverage rate for gray cast iron. Beam diameter approximately 5 mm.

12–2 WELDING

There are basically two types of laser welding, CW and pulsed. Seam welding is done both by CW CO_2 and overlapping pulses with Nd-YAG lasers. Pulsed or spot welding is done chiefly with CO_2, Nd-YAG, Nd-Glass, and, to a lesser extent, ruby lasers.

The main advantages of laser welding are

1. Minimum heat input, which results in very little distortion.
2. Small heat-affected zone (HAZ) because of the rapid cooling.
3. Narrow, generally cosmetically good, weld bead.
4. High-strength welds.
5. An easily automated process that can produce very precisely located welds.
6. Weld some metals difficult to weld by other techniques, especially dissimilar metals.
7. Weld in areas difficult to reach with some other techniques.
8. Frequently faster than other techniques.

Disadvantages of lasers in welding are

1. Extremely hard weld bead in hardenable materials, cold cracking, and hot cracking may be a problem due to rapid heating and cooling.
2. High capital investment compared with most techniques, but this factor may be more than offset by improved part quality, high laser up-time (usually over 90%), and relatively low operating and maintenance costs.

Continuous laser welding.   At present, continuous seam welding is done only with CO_2 lasers, usually 500 W or higher-power level. Somewhere around 300 to 500 W keyholing begins to play a role in continuous welding, but melting by thermal conduction is predominant up to around 1.0 kW. At multikilowatt levels, keyholing is the predominant phenomenon.

Theoretical approaches that give reasonable agreement for conduction limited welding are: simple energy balance, the uniform irradiance model, and numerical, finite difference or FEM approaches. The biggest problem with calculations in this regime is the uncertainty in the reflectance of the molten metal, which may approach that of the solid metal.

The method of Swift-Hook and Gick (see Chapter 11) works quite well for CW welding when keyholing is predominant. It is known that maximum penetration varies approximately as $P^{0.7}$, where P is power. The energy balance approach will give "ballpark" results if judicious guesses are made regarding conduction and reflection losses.

Nearly all power is absorbed in the keyhole and very efficient welds can be made at high-power levels. Weld efficiency is defined as the ratio of the energy used to melt material to the total energy input. A graphic example has been observed in the welding of the same parts at different power levels. In the welding of low carbon to mild steel, for instance, both 2.5 kW and 5 kW were used to achieve approximately 5.5-mm penetration. Parts welded at 2.5 kW could not be handled with bare hands due to heat loss from the weld zone by conduction. Parts welded at 5 kW, at about twice the speed, could be handled without discomfort.

Cover gases play an important role in welding operations and laser welding is no exception. Some unique problems, however, are associated with cover gases and their method of application in laser welding. Although not unique to CW CO_2 welding, they will be discussed at this point.

The method of application of the cover gas varies with the type of material being welded and the quality of the weld required. Figure 12–9 illustrates a method that uses a trailing cup. Gas is brought directly onto the weld through the nozzle coaxial with the beam to protect the final optical element and also is brought into the cup to provide coverage of the bead before and after the beam to prevent oxidation and/or atmospheric contamination. When full penetration is achieved, it may be necessary to provide cover for the underbead as well. Provision of an inert cover gas is especially important for highly oxidizable materials, such as aluminum and titanium.

The most frequently used cover gases are He, Ar, and N_2. Argon, because of its high atomic mass, provides the best coverage or protection against oxidation; however, it interacts with the metal vapor plasma in such a way that more laser radiation is lost in the plasma than with He. Maximum penetration depths are substantially lower for Ar and N_2 than He. Helium has a much higher thermal conductivity than either Ar or N_2 and is able to cool the plasma and reduce the beam interaction. The mechanism for the interaction is free electron absorption, for the photons at 10.6 μm are not sufficiently energetic to cause ionization or atomic excitation. The ionization potential for He is about 1.5 times that of Ar. This factor, coupled with the higher plasma temperature, results in a much greater beam interaction for Ar than He.

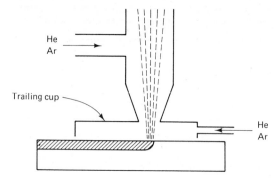

Figure 12–9 Cover gas arrangement using a trailing cup.

It is also likely that the molten pool at the surface is minimized due to the high thermal conductivity of He, thereby allowing better penetration of the beam into the keyhole. It has been observed that in high-speed welding, where maximum penetration is not achieved, deeper penetration is obtained with N_2 than He. It is possible that this situation occurs because the beam is not incident on the plasma as long and the lower thermal conductivity of N_2 allows more conduction-limited welding, combined with the fact that N_2 is better suited for dispersing the plasma because of its higher molecular mass.

Frequently the cover gas is applied by means of a tube aimed directly at the weld location if weld bead oxidation is not a serious problem. When welding materials that spatter vigorously, like rimmed steel or high-carbon steel, a strong cross flow of air is required to protect the final optical element. This cross flow must be placed sufficiently far above the weld area to avoid disturbance of the molten bead and dilution of the cover gas.

Figure 12–10 contains a micrograph of a weld bead cross section for mild steel welded to low-carbon steel. The high aspect ratio and small HAZ are evident. Figure 12–11 is a photograph of a part welded by CO_2 laser continuous welding.

Pulse welding.   Numerous materials can be pulse welded with lasers. CO_2 lasers can be used for pulse welding of small parts—so-called microwelding—but are more frequently applied to spot welding of larger parts. This process can be accomplished by electronic pulsing or simply opening and closing the shutter on CW lasers. Pulse lengths of a few tenths of seconds will yield spot welds several millimeters in diameter and weld depths of a few millimeters for multikilowatt lasers.

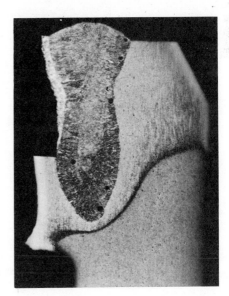

Figure 12–10   Micrograph of laser-welded part.

Figure 12–11 CO_2 laser-welded part.

The greatest bulk of pulsed laser welding is done with Nd-YAG and Nd-Glass lasers. Nd-Glass lasers are well suited for low-repetition rate spot welding on small parts, such as wires to terminals and other electronic components.

For high-speed spot welding or seam welding of such items as electronic packages, Nd-YAG lasers are much more suitable. Figure 12–12 is a photograph of a commercial Nd-YAG welder and Fig. 12–13 shows a Nd-YAG welded capacitor. Excellent hermetic welds are made by overlapping pulse welding with Nd-YAG lasers.

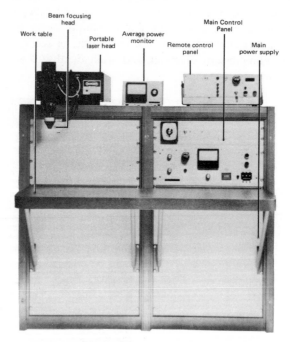

Figure 12–12 Commercial Nd-YAG welder. (Courtesy of Apollo Lasers, an Allied Company.)

Figure 12–13   Nd-YAG welded capacitor. (Courtesy of Laser Inc.)

Basically the setup for welding with Nd-YAG or Nd-Glass lasers does not differ greatly from CO_2 laser welding. Cover gases should be used whenever contamination or oxidation is a problem and a cross flow may be required to protect the optics.

Calculations for ballpark estimates of pulse times, penetration depth, and/or power requirements can be made by simple energy balance, but the graphical technique of Cohen and Epperson is superior. Typically pulse lengths are milliseconds with power densities of 10^5 W/cm².

12–3 CUTTING

Industrial laser cutting is done with CW or pulsed CO_2 and high-repetition-rate pulsed Nd-YAG lasers. The process is a gas-assist technique in which gas, under pressure, forces molten material from the kerf. Oxygen is used with oxidizable materials to increase cutting speed.

The advantages of laser cutting are essentially the same regardless of the type of laser used and may be the same as for other processing applications. These advantages include

1. Lack of tool contact
2. Ease of automation
3. Small HAZ
4. Narrow, high-precision kerf
5. Frequently higher speed than other methods
6. Ability to reach difficult-to-reach areas, including working through glass
7. Cuts low-machinability materials with relative ease

Some disadvantages are

1. High capital investment. In systems where three-dimensional contouring is required, the system cost may be ten times that of the basic laser.

2. Cut edges may be tapered and serrated when cutting thick sections.
3. The cost of oxygen or other cutting gas is substantial, although not necessarily greater than other gas-cutting techniques.

Continuous laser cutting.   Generally laser cutting is a gas-assisted process. It is possible to cut some plastics with the CO_2 laser, such as acrylic, without an assist gas, but it is desirable to use the gas assist to help control the vapors that are produced. There are nearly as many techniques for laser gas-assist cutting as there are users. All techniques, however, involve the delivery of a gas under high pressure to the point of incidence of the laser beam in such a way that the molten material is blown away. Figure 12–14 is a photograph of sheet steel being cut by a CW CO_2 laser. In this case, a coaxial laminar flow nozzle is used, which means that the beam and gas jet are coaxial. The nozzle has a special elliptical cross section leading into a straight throat, which tends to provide a columnar flow of gas on the part without turbulence. The space between the nozzle and the workpiece is about 0.5 to 1.0 mm and the pressure in the nozzle is around 35 kilopascals. In cutting oxidizable materials, such as steel or titanium, oxygen or air is used to enhance the cutting rate and quality by oxidation. Excessive oxygen, however, can cause self-burning, which will result in discouragingly poor cuts.

Some cutting systems use supersonic nozzles that tend to provide a focusing effect for the gas (similar to the exhaust plume from a rocket). Such nozzles are usually off-axis (not coaxial with the beam), although coaxial systems may be employed. Some systems use a window between the focusing optic and the

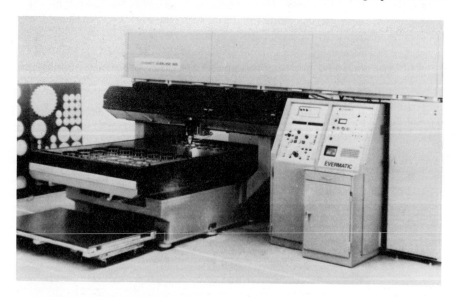

Figure 12–14   Sheet-metal cutting system. (Courtesy of Coherent Industrial Group.)

nozzle to keep the high pressure in the supersonic nozzle away from the focusing lens or mirror.

Steel around 5-mm thick is readily cut with CO_2 lasers in the 500 W range with very fine edge quality and thicknesses of low-carbon steel to 15 mm have been cut at 2.5 kW. Most metals can be cut with CW CO_2 lasers, along with a wide variety of other materials, including composites, glass, quartz, plastics, ceramics, paper, and wood. Surprisingly straight-sided kerfs can be achieved when cutting relatively thick materials as a result of a light-guiding effect due to multiple reflections from the sides of the kerf. Figure 12–15 illustrates this point with 2.5-cm-thick plastic sign letters cut at 1 kW. The same phenomenon occurs to some extent when cutting other materials. The cross section of a kerf typically looks something like the sketch in Figure 12–16.

The amount of undercutting and taper is controlled, to some degree, by the location of the focal point. In metal cutting it is usually best to focus at the surface, but for thick, nonmetallic materials it is frequently beneficial to focus below the surface. In the case of the plastic letter in Fig. 12–15, the focal point was approximately one-fourth of the thickness of the part (0.64 cm) below the surface.

The importance of maintaining good control of gas pressure and flow rate during cutting cannot be overemphasized. Accurate control of the cutting nozzle relative to the work is vital. In the case of large sheet material or parts with z-axis contours, some means of positioning the focusing column, even for very small variations in z-axis, is required. In complicated contouring four-

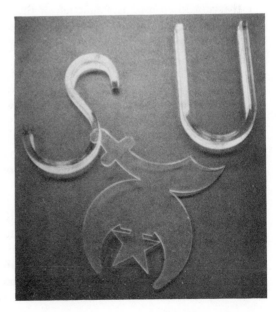

Figure 12–15 Plastic cut by CO_2 laser.

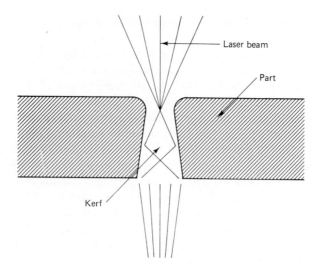

Figure 12–16   Kerf produced by laser cutting.

or five-axis numerical control with automatic location sensing and feedback control of the focusing column may be needed. In simple cases where relatively flat stock is to be cut, a two-axis numerical control system with automatic position sensing and feedback control on the z-axis may suffice. Feedback control may be accomplished through capacitive coupling to the part, air jet height measurement, and mechanical or optical position sensing. The type of sensor that should be used is a function of the nature of the part's shape, material, and surface condition. A simple technique that is effective on flat material is to "float" the focusing head mechanically by means of balls that roll on the sheet.

It is interesting to note that little, if any, damage to focusing optics due to spatter or smoke occurs during gas-assist cutting. This is true even when making blind cuts (intentionally or otherwise) because the molten material is effectively blown through the kerf or out from under the nozzle in the case of blind cuts. On the other hand, there may be a significant hazard to the operator(s) from vapors or particulates or both given off during cutting, especially plastics, some of which produce highly toxic substances when heated with a laser.

Pulsed laser cutting.   Basically there are two reasons for using pulsed laser cutting. There may simply not be sufficient power in low average power lasers to melt or vaporize the material without pulsing to achieve high peak power. For example, a Nd-YAG laser with a CW power of only a few watts, when rapidly pulsed, can be used to cut through thick film resistor material in thick film circuits to adjust resistance at a rate of several resistors per second.

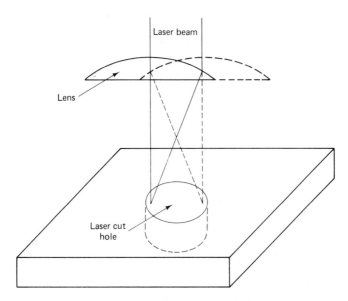

Figure 12–17 Trepanning for cutting small holes.

When cutting refractory materials, such as ceramics, it is frequently desirable to use pulsed cutting to minimize the HAZ and thereby reduce microcracking at the cut edge.

A leading application of pulsed laser cutting is the manufacture of aircraft engine parts. The low-machinability, superalloy materials used in aircraft engines, such as Waspalloy and Hastelloy-X, must be cut and drilled by EDM, ECM,[2] or laser. Both turbine blades and combustor parts are being cut and drilled by using CO_2 and Nd-YAG lasers. A laser in combination with a four- to six-axis numerical control system is a powerful and flexible tool for applications involving the cutting of thousands of holes and slots required in aircraft combustor parts. Only a program correction is required to make a design modification or a new program to process a different part.

Small holes may be cut by a technique referred to as trepanning, which is accomplished by bringing the beam off center through a rotating lens as indicated in Fig. 12–17. Such holes have been cut at angles as great as 15° to the normal to the surface.

Laser-assisted machining (LAM).   Laser-assisted machining is a process whereby a laser beam is focused on the workpiece ahead of a tool to preheat the workpiece and thus increase machinability of superalloys and ceramic materials. The tool force required for cutting is reduced and smoother cuts and higher cutting speed are possible, along with reduced tool wear. An optimum

[2] EDM is electrical discharge machining; ECM is electrochemical machining.

cutting speed must be determined for a given laser power level. Hardenable materials may harden before the tool arrives at the irradiated spot if the speed is not high enough. Hardened materials are generally easier to machine with LAM than without it.

12–4 LASER MARKING

Laser marking is a process whereby serial numbers or other identification, including logos, is placed on parts by evaporating a small amount of material with a pulsed CO_2 or Nd-YAG laser.

The advantages of laser marking are

1. High-speed, easily automated method.
2. No mechanical contact with the part.
3. Can be done through transparent enclosures and in otherwise inaccessible areas.
4. It is trivial to change patterns or index numbers.
5. Can be tied in with computer inventory control systems.

The major disadvantage is the relatively high capital investment, but it may be offset by improved inventory control and quality control.

A 10-W, average power Nd-YAG laser Q-switched by an acousto-optical modulator is capable of marking any material not transparent to the radiation. Most modern systems average over 50 W. Even transparent materials, such as acrylic, can be marked if a suitable absorbing material is placed in or on the part. The anodization can be cleanly removed from anodized aluminum, for example, leaving bright, undamaged aluminum with 8 W average power and a pulse rate of 1000 HZ. The marking or engraving speed is about 5 cm/s with the beam defocused close to 6 mm to widen the lines. Figure 12–18 is a photograph of laser-engraved, blue-anodized aluminum.

Rapid marking over a field of several centimeters can be accomplished with computer-controlled galvanometric mirrors. Incorporating part movement in the system makes field size essentially unlimited. Figure 12–19 is a photograph of a Nd-YAG laser marking system.

CO_2 lasers can be used for marking many materials transparent to 1.06-μm radiation and a few watts of average power is all that is required. Pulsing may be used, but it must be realized that electronically pulsing a CW CO_2 laser will produce powers of three to seven or eight times the CW power level. Certain types of low-power CO_2 lasers, not discussed extensively in this book, produce very high peak powers and are applicable in marking and small hole piercing applications.

One economical, fast, but less versatile technique for marking with pulsed CO_2 lasers is basically a stenciling approach. The pattern to be engraved is

Figure 12-18 Laser-engraved anodized aluminum.

defined in a reflective masking material, such as copper, and the entire mask is irradiated and an image is optically produced on the part. If deep engraving is to be done, as in laser wood engraving, the beam can be scanned across a full size mask in a raster scan pattern.

12-5 HOLE PIERCING

Laser hole piercing, frequently referred to as drilling for short, is chiefly done with pulsed CO_2 and Nd-YAG, although ruby and Nd-Glass lasers are some-

Figure 12-19 Nd-YAG laser marking system. (Courtesy of General Photonics Corp.)

times applied. The applications for laser hole piercing are incredibly varied, ranging from piercing holes in cigarette filter paper, baby bottle nipples, and aerosol can nozzles to diamonds and turbine blades for aircraft engines. Hole piercing rates vary from millions of holes per minute for paper to several seconds per hole in turbine blades.

The advantages of hole piercing with lasers are essentially the same as for cutting or welding. It is a technique in which highly reproducible holes can be made rapidly—frequently in materials of low machinability—with no tool contact.

Yet some important disadvantages to laser drilling should be considered. Laser-drilled holes do not have straight, smooth walls. A sketch of a typical laser drilled in a metal hole is given in Fig. 12–20. In metals dross may be attached to the underlip of the hole and recast metal may protrude at the top. The material is rarely completely vaporized in metal hole drilling. Instead, some metal is vaporized and up to 90% is melted and percussively removed from the hole. When a gas assist is used, the molten material is blown from the hole by the gas assist. Molten metal tends to flow up the walls under the influence of the vapor and some resolidifies on the rim of the hole. In addition, molten material may not be completely ejected from the bottom of the hole, thus resulting in dross. The dross is usually easily removed by shot blasting or light grinding, but the recast material on the top is hard and an integral part of the remaining metal. Proper techniques can minimize or eliminate both dross and recast metal on the top surface.

Hole taper is an unavoidable problem associated with beam focusing in fairly high aspect ratio holes. It, too, can be minimized by proper focusing. In multiple shot drilling the focal point can be adjusted between shots to control hole shape.

Laser hole piercing in metals has its greatest use in applications in which numerous holes must be precisely located in a highly reproducible manner. The holes routinely placed in turbine blades and turbine combustor parts for

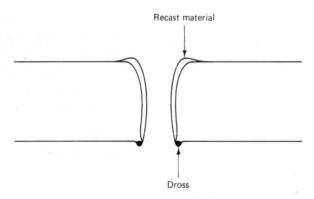

Recast material

Dross

Figure 12–20   Sketch of laser-drilled hole.

airflow control are an excellent example of this type of application, as can be seen in Fig. 12–21.

Estimates of hole-piercing parameters can be made on the basis of simple energy balance and work quite well for vaporization drilling. The pulse time required to drill a given thickness hole can be estimated by the technique outlined in Chapter 11 for a propagating vapor front. As noted, however, the development of a metal vapor plume will seriously attenuate the laser power reaching the workpiece. Consequently, pulses must be kept short enough to allow the developing plume to clear the hole. A gas assist or cross flow or both will help reduce this problem. Typical pulse lengths are tenths of milliseconds at power densities on the order tens to hundreds of megawatts.

Holes are routinely laser drilled in ceramic materials. For electrical connections, holes are drilled completely through the boards. Ceramic sheets are scribed by laser drilling a row of blind holes and then mechanically snapping the sheet into two or more parts. Both pulsed CO_2 and Nd-YAG can be used for ceramic drilling, although CO_2 is faster. Pulses should be kept short to maximize vaporization and decrease heat loss by conduction, which produces thermal stresses and hence cracking. Also, rapid pulses minimize the amount of resolidified material in and around the hole, which reduces hole quality. Typically pulse lengths are one to a few milliseconds with power levels of 50 W for CO_2 lasers and pulse energy of a few joules for Nd-YAG lasers.

It has been found that the nature of Nd-YAG laser pulses can seriously affect hole quality [Roos 1980]. Normal relaxation pulses, which consist of a series of initial spikes followed by continuous output, can result in clogged holes due to the material melted by the continuous portion of the pulse. A train of spikes produced by loss modulation provides superior holes in thin materials drilled at relatively low energy per pulse (200 mJ or less per pulse). Proper application of a gas assist also minimizes this problem.

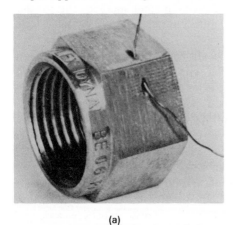

(a)   (b)

Figure 12–21   Laser-drilled holes. (Courtesy of Laser Inc.)

Other materials that are drilled with lasers are rubber and many plastics drilled with CO_2 lasers and diamonds and other gemstones drilled with Nd-YAG lasers.

12-6 ALLOYING AND CLADDING

Two techniques that show great promise for metal surface modification to improve wear, abrasion, corrosion, impact resistance, and strength while retaining desirable substrate metal properties are laser alloying and cladding. In both cases, an alloy material, usually a powder, is placed on a metallic surface and melted with a CO_2 laser beam. In alloying a substantial portion of the substrate material is melted and mixed with the melted powder, thus producing a true alloy layer on the surface after solidification. In cladding only enough of the substrate is melted to produce a metallurgical bond between the alloy material and the substrate surface. This process is also called *hardfacing*. The technique is similar to heat treatment in that a defocused or integrated CO_2 laser beam is traversed over the region to be melted.

Because the spot size is large (~1.0 cm) and melt depths are only on the order of a millimeter, reasonable calculations can be made by using the uniform irradiance model or any of the more sophisticated models used in heat treatment analysis; however, the heat of fusion should be taken into account.

Some materials used in this work are cobalt or nickel-based chromium carbides and tungsten carbide nickel alloys. Methods for applying the powders prior to laser melting include powder-solvent slurries, mixing the powder with an organic binder, and plasma spraying. The biggest problem involves applying the powder with the appropriate thickness and applying it consistently. Mechanical scrapers can be used to remove excess powder to a given thickness. An automated spraying technique may also be used. In some methods the powder is sprayed through a small orifice and simultaneously melted onto the surface by the laser beam.

The advantages of laser cladding and alloying are similar to those found in laser heat treatment and surface melting with the additional advantages of reduced alloy material consumption and greater control and uniformity of melt depth. The dilution of the cladding by base metal is negligible in cladding. In alloying extremely good mixing of the alloying material and substrate metal occurs due to a vigorous mixing action that takes place in the melt pool.

Surface tension and thermal gradient effects produce strong convection currents in the melt puddle. These currents rise at the point of incidence of the beam and flow outwards toward the walls and down in a circulatory manner. This mixing does cause rippling in the solidified material, both in the direction of the track and transverse to it. Generally the region of overlap between adjacent tracks will not be flat either. So if a smooth, flat surface is required, a final finishing operation is required.

12-7 MISCELLANEOUS APPLICATIONS

There are many other material processing applications of lasers. Some notable ones are soldering and desoldering with CO_2 and Nd-YAG lasers, wire insulation stripping with CO_2 lasers, and cutting of silicon for solar cells with Nd-YAG lasers.

PROBLEMS

12-1. Describe, with the aid of sketches, the process of laser surface hardening. Explain what the important parameters are and discuss the phenomena of hardening, quenching, and similar processes.

12-2. Describe the process of laser welding. Explain the use of cover gases and other important parameters. Briefly discuss both CW and pulse welding with respect to the type of laser used and the power ranges involved. Explain what keyholing is, what causes it, and point out what power range it occurs in.

12-3. Describe and discuss pulsed and CW laser cutting and point out what lasers are used for various types of materials.

12-4. Discuss and describe hole piercing with lasers.

12-5. Discuss and describe the processes of alloying and cladding with lasers.

12-6. Discuss the advantages and disadvantages of using lasers in the processes mentioned in Problems 12-1 through 12-5.

REFERENCE

Roos, Sven-Olov, "Laser Drilling with Different Pulse Shapes," *J. Appl. Phys.*, September 1980, pp. 5061–5063.

BIBLIOGRAPHY

12-1. Charschan, S. S. (ed.), *Lasers in Industry*. New York: Van Nostrand Reinhold Co., 1972.

12-2. Ready, J. F., *Industrial Applications of Lasers*. New York: Academic Press, 1978.

12-3. Duley, W. W., *CO_2 Lasers: Effects and Applications*. New York: Academic Press, 1976.

12-4. Ferris, S. D., H. J. Leamy, and J. M. Poate (eds.), "Laser-Solid Interactions and Laser Processing—1978," *AIP Conference Proceedings No. 50*. New York: American Institute of Physics, 1979.

12-5. Ready, J. F. (ed.), *Lasers in Modern Industry*. Dearborn, MI: Society of Manufacturing Engineers, 1979.

12–6. Metzbower, E. A. (ed.), *Applications of Lasers in Materials Processing*. Metals Park, OH: American Society for Metals, 1979.

12–7. Engineering Staff of *Coherent, Lasers: Operation, Equipment, Application and Design*. New York: McGraw-Hill Book Co., 1980.

12–8. Ready, J. F. (ed.), "Industrial Applications of High Power Laser Technology," *Proceedings of the Society of Photo-Optical Instrumentation Engineers*, Vol. 86, Bellingham, WA, SPIE, 1977.

12–9. Ready, J. F. (ed.), "Laser Applications in Materials Processing," *Proceedings of the Society of Photo-Optical Instrumentation Engineers*, Vol. 198, Bellingham, WA, SPIE, 1980.

12–10. Laser Institute of America, *Use of Lasers in Materials Processing Applications*. Toledo, OH, LIA.

12–11. Proceedings of the First International Laser Processing Conference. Toledo, OH: Laser Institute of America, 1981.

12–12. Ream, S. L., "High-Power Laser Processing," 1981 SME Heat Treating Conference, MR81–316. Dearborn, MI: Society of Manufacturing Engineers, 1981.

12–13. Gnanamuthu, D. S., "Laser Surface Treatment," *Optical Engineering*, **19**, No. 5, September–October 1980, 783–792.

12–14. Bolin, S. R., "Bright Spot for Pulsed Lasers," *Welding Design and Fabrication*, August 1976, pp. 74–76.

12–15. Miller, F. R., "Advanced Joining Processes," *SAMPE Quarterly*, October 1976, pp. 46–54.

12–16. Banas, C. M., "High Power Laser Welding—1978," *Optical Engineering*, May–June, 1978, pp. 210–219.

12–17. Bolin, S. R., "Part 2: Laser Welding, Cutting and Drilling," *Assembly Engineering*, July 1980, pp. 24–27.

12–18. Benedict, G. F., "Production Laser Cutting of Gas Turbine Components," *Society for Manufacturing Engineers Paper No. MR80–851*, Dearborn, MI, 1980.

12–19. Engel, S. L., "Laser Cutting of Thin Materials," Society of Manufacturing Engineers Paper No. MR74–960, 1974.

12–20. Terrell, N. E., "Laser Precision Hole Drilling," *Manufacturing Engineering*, May 1982, pp. 76–78.

12–21. LIA Laser-Material Processing Committee, "Laser Drilling," *Electro-Optical Systems Design*, July 1977, pp. 44–48.

12–22. Bolin, S., "Laser Drilling and Cutting," Society of Manufacturing Engineers Paper No. MR81–365, 1981.

12–23. Jau, B. M., S. M. Copley, and M. Bass, "Laser Assisted Machining," Society of Manufacturing Engineers Paper No. MR80–846, 1980.

12–24. Jacobson, D. C., et al., "Analysis of Laser Alloyed Surfaces," *IEEE Transactions on Nuclear Science*, **NS-28**, No. 2, April 1981, 1828–1830.

12–25. Bass, M. (ed.), *Laser Materials Processing*. New York: North-Holland Publishing Co., 1983.

Index